Physics and Chemistry in Space
Volume 5

Edited by
J. G. Roederer, Denver

A. J. Hundhausen

Coronal Expansion and Solar Wind

With 101 Figures

Springer-Verlag Berlin Heidelberg New York 1972

A. J. Hundhausen

High Altitude Observatory, National Center for Atmospheric Research*,
Boulder, CO/USA

* The National Center for Atmospheric Research is sponsored by the National Science
Foundation.

ISBN-13:978-3-642-65416-9   e-ISBN-13:978-3-642-65414-5
DOI: 10.1007/978-3-642-65414-5

A. J. Hundhausen

Coronal Expansion and Solar Wind

With 101 Figures

Springer-Verlag New York Heidelberg Berlin 1972

A. J. Hundhausen

High Altitude Observatory, National Center for Atmospheric Research*,
Boulder, CO/USA

* The National Center for Atmospheric Research is sponsored by the National Science
Foundation.

ISBN-13:978-3-642-65416-9   e-ISBN-13:978-3-642-65414-5
DOI: 10.1007/978-3-642-65414-5

Dedicated to My Parents

Preface

Little more than ten years have passed since spaceprobe-borne instruments conclusively demonstrated the existence of the solar wind. These observations confirmed the basic validity of a theoretical model, first proposed by E. N. Parker, predicting a continuous, rapid expansion of the solar corona. The subsequent decade has seen a tremendous growth in both the breadth and sophistication of solar wind observations; the properties of the interplanetary plasma near the orbit of the earth are now known in great detail. The theory of the coronal expansion has also been highly refined both in the sense of including additional physical processes, and of treating more realistic (time-dependent and non spherically-symmetric) coronal boundary conditions.

The present volume is an attempt to synthesize the solar wind observations and coronal expansion models from this decade of rapid development. The ultimate goal is, of course, the interpretation of observed solar wind phenomena as the effects of basic physical processes occurring in the coronal and interplanetary plasma and as the natural manifestations of solar properties and structures. This approach implies an emphasis upon the "large-scale" features revealed by the observations. It requires extensive use of the concepts and methods of fluid mechanics.

It is the author's intent that this presentation be intelligible to the advanced graduate student or the more experienced scientist unfamiliar with its specialized subject matter. Essential to understanding of the treatment is a grasp of the basic conservation laws of fluid and particle mechanics. Some familiarity with the terminology of solar physics (in particular that describing solar activity) is assumed. Mathematical developments will be held to a minimum.

Gaussian—c.g.s. units will be used with two exceptions dictated by prevailing conventions in this field of research. Almost all solar wind observers give solar wind speeds in $km\ sec^{-1}$ and interplanetary magnetic flux densities $|B|$ in units of $\gamma = 10^{-5}$ gauss.

The author's viewpoints regarding the material presented herein were largely formed during his association with the Vela satellite group at the Los Alamos Scientific Laboratory. The collaboration and scientific stimulation of Drs. S. J. Bame, M. D. Montgomery, and R. A. Gentry of that institution were of incalculable value to the author. Among the influences exerted by colleagues from other institutions, those of Drs. A. Barnes, L. F. Burlaga, D. S. DeYoung, J. T. Gosling, J. Hirshberg, N. F. Ness, K. H. Schatten, and J. M. Wilcox have had the greatest impact upon the ideas set forth here.

Special acknowledgements are due to Drs. S. J. Bame and A. J. Lazarus for permission to use unpublished data. The manuscript was typed with diligence and accuracy by Mrs. Ruby Fulk. Its reading by Dr. J. T. Gosling produced abundant and detailed criticisms to the author's benefit. Final credit must go to my wife Joan, not only for her patience during the struggle, but for application of the mathematical and grammatical reins.

October 1972   A. J. Hundhausen

Contents

Chapter IV

Chemical Composition of the Expanding Coronal and Interplanetary Plasma

Chapter V

High-Speed Plasma Streams and Magnetic Sectors

Chapter VI

Flare-Produced Interplanetary Shock Waves

Chapter VII

Concluding Remarks

Chapter I

History and Background

I.1 Introduction

The solar wind plays an important and unifying role in modern space research, commanding attention as much for its relevance to many solar, geophysical, and astronomical phenomena as for its intrinsic physical interest. In the light of this role, it may be difficult to recall that the very existence of this continuous flow of solar material through interplanetary space, as well as a particular model of its origin in the hydrodynamic expansion of the solar corona, was hotly debated less than fifteen years ago. The universal acceptance now accorded these concepts was achieved only after the advent of *in situ* interplanetary observations in the early 1960's. Although our focus in this volume will be on recent observations, interpretations, and models of solar wind phenomena, a brief review of earlier developments will serve both to place the recent work in perspective and to introduce some of the physical background basic to the subsequent discussion.

I.2 Indirect Evidence for the Existence of the Solar Wind

The study of solar terrestrial relationships, pursued extensively during the first half of this century, provided considerable evidence for the presence of solar material in interplanetary space. "Solar corpuscular radiation" was widely invoked to explain polar aurorae, geomagnetic activity, and cosmic ray modulations. A quantitative model describing the impact of a solar flare-produced cloud of ionized gas upon the geomagnetic field was developed in the 1930's by Chapman and Ferraro [1.1, 1.2] to explain the sudden commencement and subsequent phases of geomagnetic storms. Some of the ideas evolved during this era are remarkably close to those in vogue today; in particular, the model of Chapman and Ferraro remains the basis for our present day understanding of the solar wind—geomagnetic interaction. However, most of these ideas concerned the effects of solar activity, and the postulated solar particle emissions were generally held to be of a transient, rather than a continuous, nature.

In fact, some evidence for continuous solar corpuscular radiation can be found in these same examples of solar-terrestrial relationships [1.3]. Aurorae are almost always observed in the polar regions, the geomagnetic field is never completely steady, and cosmic rays show a solar-oriented modulation. However, other explanations could be proposed for each of these effects, and it is largely in retrospect that they emerge as clear evidence for the continuous transport of solar material to the orbit of the earth. The concept of interplanetary space as a slightly dusty vacuum, only occasionally disturbed by streams of solar particles or plasma, prevailed for the first half of the century.

Two new pieces of observational evidence for the continuous presence of an interplanetary gas were proposed in the 1950's. First, the zodiacal light, traditionally attributed to the scattering of sunlight by interplanetary dust particles, was found to be strongly polarized. Arguing that such an effect would not be produced by scattering from dust, Behr and Siedentopf [1.4] suggested that an important contribution to the zodiacal light was produced by scattering from electrons with a density of $\sim 10^3 \mathrm{cm}^{-3}$ near the earth. Second, features of ionic comet tails, whose anti-solar orientation (independent of the direction of orbital motion of the cometary nucleus) had been traditionally attributed to the pressure of solar radiation, were found to undergo unaccountably large accelerations. Biermann [1.5, 1.6] suggested that the acceleration (and ionization) of cometary molecules was due to interaction with an interplanetary background of ions flowing continuously and radially from the sun. On the basis of a simple model of this interaction, Biermann found that a flux density (near the orbit of the earth) of $\sim 10^{10} \mathrm{ions\,cm}^{-2} \mathrm{sec}^{-1}$ was required to explain the observed accelerations. Combination of this flux value with the $10^3 \mathrm{cm}^{-3}$ density inferred by Behr and Siedentopf led to an estimated ion speed of $\sim 10^7 \mathrm{cm\,sec}^{-1}$.

Neither the electron scattering explanation of zodiacal light polarization nor the ionic source of comet tail ionization and acceleration are accepted today. Scattering from particulate matter is held to produce the zodiacal light (and its observed polarization), and spacecraft observations made in interplanetary space yield electron densities two orders of magnitude lower than inferred from the electron scattering interpretation. The solar wind-comet interaction is known to be far different from that proposed by Biermann (although the orientations and accelerations of ionic comet tails do stem from that interaction) and the directly observed ion flux densities are two orders of magnitude lower than inferred from the original model. Nonetheless, the zodiacal light and comet tail studies of the early 1950's played key roles in the development of the solar wind concept. These studies led to careful theoretical study of the dynamics of the solar corona and to the discovery that a continuous, rapid expansion into interplanetary space was a natural consequence of the high coronal temperature.

I.3 The Extension of the Solar Corona into Interplanetary Space

By 1950, several lines of observatonial evidence had led to the recognition that temperatures of $\sim 10^6\,°$K were characteristic of the solar corona. As the ionization of hydrogen is nearly complete at such temperatures, the corona is expected to be a proton-electron gas with a small admixture of ions of other (much less abundant) elements. The electron component of this gas scatters the photospheric radiation to produce the "white light" corona observed during total solar eclipses to extend above the photosphere. The intensity of this scattered light implies an electron density of 10^8 to $10^9\,\mathrm{cm}^{-3}$ near the base of the corona, and an outward decrease with a scale height of ~ 0.1 solar radii.

The thermal conductivity, κ, of a proton-electron gas is primarily due to the more mobile electrons, and is given by [1.7]

$$\kappa = \kappa_0\, T_e^{\frac{5}{2}} \tag{1.1}$$

where T_e is the electron temperature and κ_0 is a weak function of density and temperature. For typical coronal conditions $\kappa_0 \approx 8 \times 10^{-7}\,\mathrm{ergs\,cm}^{-1}$ $\sec^{-1}\deg^{-7/2}$. In the presence of a temperature gradient (and neglecting the effects of any magnetic field) the coronal plasma is expected to conduct heat at the rate

$$f_e = -\kappa \nabla T_e\,.$$

At prevailing coronal temperatures, the electron conductivity is extremely high; $T_e = 10^6\,°$K gives $\kappa = 8 \times 10^8\,\mathrm{ergs\,cm}^{-1}\sec^{-1}\deg^{-1}$, about twenty times the conductivity of copper at room temperature.

The implications of this high conductivity regarding the structure of the corona were explored by Chapman [1.8] in 1957. For a spherically-symmetric, static corona, the absence of any energy sources or sinks reduces the heat transfer equation to

$$-\nabla \cdot f_e = \frac{1}{r^2}\frac{d}{dr}\left(r^2 \kappa_0\, T_e^{\frac{5}{2}}\frac{dT_e}{dr}\right) = 0 \tag{1.2}$$

where r is the heliocentric distance. Under the boundary condition that the temperature must approach zero at large distances from the sun, (1.2) has the solution

$$T_e = T_{e0}\left(\frac{r_0}{r}\right)^{\frac{2}{7}} \tag{1.3}$$

where $T_{e0}=T_e(r_0)$ and r_0 is some reference level. If r_0 is taken to be 7.36×10^{10} cm (1.0575 solar radii) and T_{e0} to be 10^6 °K, the electron temperature at the orbit of the earth, $r_e = 1.496 \times 10^{13}$ cm (hereafter referred to as the astronomical unit, or 1 AU), is 2.19×10^5 °K. The high conductivity of coronal material implies a small temperature gradient, and thus an extension of the high coronal temperature well out into interplanetary space.

In static equilibrium, the downward force of gravity upon any fluid parcel is balanced by the upward pressure force, or

$$\frac{dP}{dr} = -\frac{GM_\odot \rho}{r^2} \tag{1.4}$$

where P is the pressure and ρ the mass density of the coronal gas, G is the gravitational constant, and $M_\odot$ is the solar mass. The electrostatic forces between electrons and protons require nearly equal number densities on all but the smallest distance scales (smaller than the Debye length of the plasma), so that

$$\rho = n(m_e + m_p) \approx nm,$$

where n is the density of either constituent, m_e and m_p are the electron and proton masses and m is the mass of a hydrogen atom. If a single temperature T is assumed for both electrons and protons, the pressure is given by

$$P = 2nkT.$$

Use of solution (1.3) for the temperature in the hydrostatic equilibrium equation (1.4) gives

$$\frac{d}{dr}\left(\frac{n}{r^{\frac{2}{7}}}\right) = -\frac{GM_\odot m}{2kT_0 r_0^{\frac{2}{7}}} \frac{n}{r^2}.$$

A solution is

$$n(r) = n_0 \left(\frac{r}{r_0}\right)^{\frac{2}{7}} \exp\left\{\frac{7}{5}\frac{GM_\odot m}{2kT_0 r_0}\left[\left(\frac{r_0}{r}\right)^{\frac{5}{7}}-1\right]\right\}, \tag{1.5}$$

where $n_0 = n(r_0)$. For r near r_0, (1.5) reduces to

$$n(r) \approx n_0 \exp\left\{-\frac{GM_\odot m}{2kT_0 r_0}\frac{r-r_0}{r_0}\right\},$$

or to the standard expression for a "thin" atmospheric shell with scale height $2kT_0 r_0^2/GM_\odot m$.

A surprising implication of the solution (1.5) for a static corona is the high density value near the earth. Reasonable choices of the coronal density n_0 and temperature T_0 give electron densities of 10^2 to 10^3 cm^{-3} at $r=1$ AU. In other words, the extension of the high coronal temperature into interplanetary space leads to large scale heights (or small density gradients) at large heliocentric distances (where the gravitational force has become small), and hence to large interplanetary electron densities. In fact, the near-earth density so derived by Chapman agreed with that deduced from the zodiacal light interpretation of Behr and Siedentopf, and perhaps lent some credence to that interpretation. The Chapman model itself leads to a fundamental difficulty at large heliocentric distances (as will be explored in I.4). Nonetheless, its importance should not be underestimated, as this model demonstrated that the corona could not terminate near the sun; rather coronal (and thus solar) material must extend far out into interplanetary space.

I.4 The Expansion of the Solar Corona into Interplanetary Space

The pressure $P(r)$ implied by the Chapman model of a static, conducting corona is obtained by combining equations (1.3) and (1.5):

$$P(r) = P_0 \exp\left\{ \frac{7}{5} \frac{GM_\odot m}{2kT_0 r_0} \left[\left(\frac{r_0}{r}\right)^{\frac{5}{7}} - 1 \right] \right\}, \qquad (1.6)$$

where $P_0 = 2n_0 k T_0$. Thus $P(r)$ decreases monotonically with increasing r, approaching the finite value $P_\infty = P_0 \exp(-(7/5)(GM_\odot m/2kT_0 r_0))$ as $r \to \infty$. This finite pressure at larger r is a direct consequence of the density variation given by (1.5). The density attains a minimum value at $r = ((7/4)(GM_\odot m/r_0 k T_0))^{7/5} r_0$ and increases beyond this heliocentric distance, ultimately varying as $r^{2/7}$.

For coronal densities and temperatures consistent with observed values, the Chapman model gives a pressure of about 10^{-5} dynes cm^{-2} at large heliocentric distances. One must expect that the extended corona should merge into the interstellar background as $r \to \infty$. Physical conditions in the interstellar medium are poorly known, even today, but the combined pressure of the galactic magnetic field, interstellar gas, and cosmic rays is thought to be 10^{-12} to 10^{-13} dynes cm^{-2}. The Chapman coronal model cannot blend into such a background, as the mismatch in pressure is seven or eight orders of magnitude. The implication of this mismatch was realized by Parker [1.9], who argued that such a model could be a valid representation of the corona only if an unreasonably large pressure were to exist in interstellar space. In the absence of such a pressure, Parker concluded that "... probably it is not possible for the solar corona, or, indeed, perhaps the atmosphere of any star, to be in complete hydrostatic equilibrium out to large heliocentric distances." But if hydrostatic equilibrium were not possible, what could be the equilibrium configuration of the corona? Apparently motivated by

Biermann's arguments for a continuous, rapid outflow of material in interplanetary space, Parker considered a new theoretical possibility, a corona undergoing a steady expansion.

Let us then consider a spherically symmetric corona in steady motion, so that all physical properties depend upon heliocentric position r, but not upon time. The equation of hydrostatic equilibrium (1.4) in Chapman's treatment must then be replaced by the equation for momentum conservation in a fluid,

$$\rho u \frac{du}{dr} = -\frac{dP}{dr} - \rho \frac{GM_\odot}{r^2} \tag{1.7}$$

where u is the speed of expansion, assumed to be purely radial. To this momentum equation must be added an equation of mass conservation

$$\frac{1}{r^2} \frac{d}{dr} (r^2 \rho u) = 0 \tag{1.8}$$

and an equation of energy conservation

$$\frac{1}{r^2} \frac{d}{dr} \left[r^2 \rho u \left(\frac{1}{2} u^2 + \frac{3}{2} \frac{P}{\rho} \right) \right] = -\frac{1}{r^2} \frac{d}{dr} (r^2 P u) - \rho u \frac{GM_\odot}{r^2} + S(r). \tag{1.9}$$

The term $S(r)$ represents any energy source or sink (e. g., radiation, heat conduction) other than those associated with the body forces included in the momentum equation. Correspondence with the Chapman model is retained by considering only the energy source due to heat conduction,

$$S(r) = -\frac{1}{r^2} \frac{d}{dr} (r^2 f_c),$$

where f_c is the radial heat conduction flux density. If we again regard the corona as an electrically neutral proton-electron gas, the mass density is

$$\rho = n(m_p + m_e) \approx nm$$

and the pressure is given by the equation of state

$$P = nk(T_e + T_p),$$

where T_e and T_p are respectively the electron and proton temperatures.

The simultaneous solution of the three nonlinear conservation equations (1.7), (1.8), and (1.9) is a difficult task and will be discussed extensively in Chapter III. In his initial formulation of this problem, Parker introduced the simplifying assumption that pressure and density are related by a so-called polytropic law,

$$P = P_0 \left(\frac{\rho}{\rho_0} \right)^\alpha, \tag{1.10}$$

where α is a constant, the so-called "polytropic index". This effectively assumes a solution of the energy equation (1.9) (with an implied source term that may not correspond to physical reality), eliminating that equation from the system. In describing Parker's treatment of the expanding corona, we will limit our detailed discussion to a single limiting case, that of $\alpha=1$, corresponding to an isothermal corona (or, more precisely, to a constant sum of the electron and proton temperatures). The equation of state is then

$$P=2nkT$$

where $T=\frac{1}{2}(T_e+T_p)$ is constant.

The mass and momentum conservation equations can be written as

$$\frac{1}{r^2}\frac{d}{dr}(r^2nu)=0 \tag{1.11}$$

and

$$nmu\frac{du}{dr}=-2kT\frac{dn}{dr}-nm\frac{GM_\odot}{r^2}. \tag{1.12}$$

Equation (1.11) has the first integral

$$4\pi nur^2=I, \quad \text{a constant,} \tag{1.13}$$

that simply states the constancy of the proton (or electron) flux through any sun-centered spherical surface. Using (1.13) to eliminate the number density n from (1.12) gives

$$\frac{1}{u}\frac{du}{dr}\left(u^2-\frac{2kT}{m}\right)=\frac{4kT}{mr}-\frac{GM_\odot}{r^2}. \tag{1.14}$$

Let us restrict attention to temperatures $T<GM_\odot m/4kr_0$, where r_0 is the inner boundary of the semi-infinite range $r_0\leq r<\infty$ over which (1.14) applies (i.e., r_0 is the "base" of our model corona). The right-hand side of (1.14) is then negative for $r_0<r<r_c$, where

$$r_c=\frac{GM_\odot m}{4kT} \tag{1.15}$$

($r_c>r_0$ because of the temperature restriction assumed above), and is positive for $r_c<r<\infty$. The zero at $r=r_c$ implies that the left-hand side of (1.14) must also be zero at this position, the so-called "critical radius." This can occur in two ways; either because

$$u^2(r_c)=u_c^2=\frac{2kT}{m} \tag{1.16}$$

or because

$$\frac{1}{u}\frac{du}{dr}\bigg|_{r=r_c}=0. \tag{1.17}$$

We are interested in single-valued solutions for which $u(r)$ and du/dr are continuous. If (1.16) is satisfied, then du/dr has the same sign for all r; $u(r)$ is either monotonically increasing or monotonically decreasing. If (1.17) is satisfied, then $u^2 - 2kT/m$ has the same sign for all r; $u(r)$ has a maximum near $r = r_c$ if $u^2(r_c) < 2kT/m$, or a minimum near $r = r_c$ if $u^2(r_c) > 2kT/m$. These possibilities lead to the existence of four different classes of solutions to equation (1.14), with the properties illustrated in Fig. 1.1.

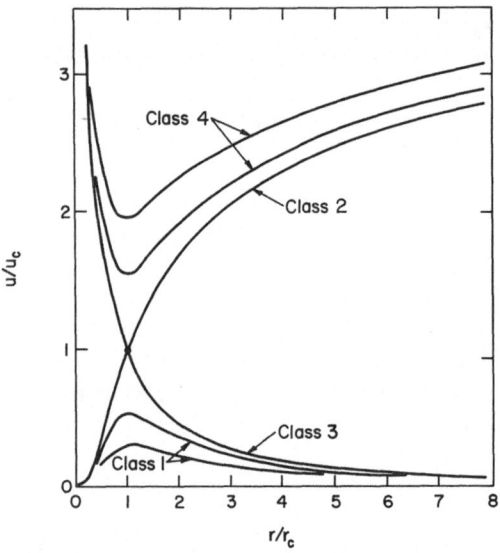

Fig. 1.1   The topology of solutions to equation (1.14). The critical radius r_c and expansion speed u_c are defined in the text

Class 1, a family of solutions for which $u(r)$ increases with r for $r_0 \leq r \leq r_c$, attains a maximum value less than $(2kT/m)^{\frac{1}{2}}$ near $r = r_c$ and decreases with r for $r_c < r < \infty$.
Class 2, a unique solution for which $u(r)$ is monotonically increasing, with $u^2(r_c) = 2kT/m$.
Class 3, a unique solution for which $u(r)$ is monotonically decreasing, with $u^2(r_c) = 2kT/m$.
Class 4, a family of solutions for which $u(r)$ decreases with r for $r_0 \leq r < r_c$, attains a minimum value greater than $(2kT/m)^{\frac{1}{2}}$ near $r = r_c$, and increases with r for $r_c < r < \infty$.

Each of the solution classes given above fits a different set of boundary conditions at $r = r_0$ and $r \to \infty$. The *physical* acceptability of these solutions depends upon these boundary conditions. For example, both

Class 3 and Class 4 can be ruled out as plausible models for the solar corona as all solutions in these classes have $u(r) > 2kT/m$ for r near r_0. With $T \approx 10^6$ °K, this implies expansion speeds at $\sim 10^7$ cm sec^{-1} deep in the corona; such large coronal motions are not observed. Both Class 1 and Class 2 remain acceptable models for the solar corona on the basis of their properties near r_0, as all solutions in these classes have small expansion speeds near r_0. However, the Class 1 and Class 2 solutions display entirely different behaviors as $r \to \infty$, and the acceptability of these two classes hinges on this difference.

The behavior of solutions to (1.14) as $r \to \infty$ is easily understood if the derivative du/dr is rewritten as $(1/2u)(du^2/dr)$ to give

$$\frac{1}{2} \frac{du^2}{dr} \left(1 - \frac{2kT}{m} \frac{1}{u^2} \right) = \frac{4kT}{mr} - \frac{GM_\odot}{r^2}. \tag{1.18}$$

A solution of (1.18) is then

$$\frac{1}{2} u^2 - \frac{kT}{m} \ln u^2 = \frac{4kT}{m} \ln r + \frac{GM_\odot}{r} + C \tag{1.19}$$

where C is a constant of integration. This solution can be expressed in the more convenient form

$$\left(\frac{u}{u_c} \right)^2 - \ln \left(\frac{u}{u_c} \right)^2 = 4 \ln r + \frac{GM_\odot m}{kTr} + C' + \ln u_c^2 \tag{1.20}$$

where $u_c^2 = 2kT/m$ as in (1.16). For the family of solutions in Class 1, u/u_c is less than one and decreasing as $r \to \infty$. Hence $(u/u_c)^2 \ll |-\ln(u/u_c)^2|$ and (1.20) becomes

$$\ln \frac{u}{u_c} \approx -2 \ln r$$

so that

$$u \propto \frac{1}{r^2}.$$

The flux integral (1.13) then implies that the density n must approach a constant, finite value n_∞ as $r \to \infty$. Thus the Class 1 solutions would again give finite pressures $P = 2n_\infty kT$ at large r, and would lead to the same difficulty in merging into the interstellar background as encountered in Chapman's static solution. For the unique solution of Class 2, u/u_c is greater than one and increasing as $r \to \infty$. Then $(u/u_c)^2 \gg |-\ln(u/u_c)^2|$ and (1.20) becomes

$$\left(\frac{u}{u_c} \right)^2 \approx 4 \ln r$$

so that

$$u \approx 2u_c (\ln r)^{\frac{1}{2}}$$

or the expansion speed continues to increase slowly as $r \to \infty$. The flux integral (1.13) then implies that $n \to 0$ as $r \to \infty$. Thus the Class 2 solution gives $P \to 0$ at large r, and could merge into the low pressure, interstellar background.

Parker's consideration of the equations of motion for an expanding isothermal corona thus revealed the existence of a unique solution for which $u(r)$ increased from a low value, reconcilable with coronal observations, at small heliocentric distances to a large expansion speed and vanishing pressure, the latter reconcilable with knowledge of the interstellar medium, at large heliocentric distances. Evaluation of the integration constant C' in (1.20) leads to the explicit expression for this solution

$$(u^2 - u_c^2) - u_c^2 \ln\left(\frac{u^2}{u_c^2}\right) = 4 u_c^2 \ln\left(\frac{r}{r_c}\right) + 2 G M_\odot \left(\frac{1}{r} - \frac{1}{r_c}\right). \qquad (1.21)$$

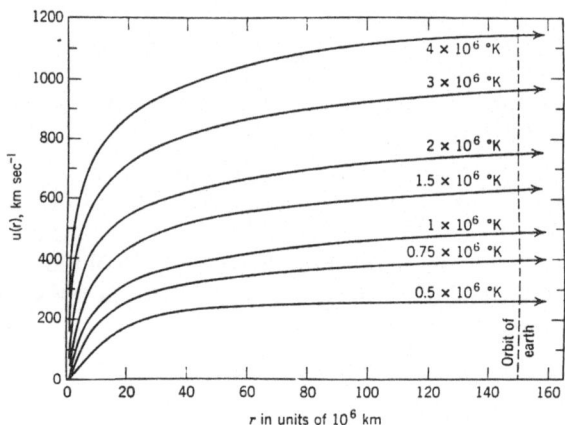

Fig. 1.2   Expansion speed as a function of heliocentric distance for isothermal coronas of various temperatures [1.10]

Fig. 1.2 shows the resulting $u(r)$ for different coronal temperatures and with $r_0 = 10^{11}$ cm (or 1.4 solar radii). Reasonable values of T lead to expansion speeds of several times 10^7 cm sec^{-1} in the interplanetary region. Such flow speeds are, of course, in agreement with those inferred by Biermann. As the sound speed in the expanding plasma is $c_s = (\gamma P/\rho)^{\frac{1}{2}} = \gamma^{\frac{1}{2}} u_c$, where γ is the ratio of specific heats, solutions of the form (1.21) predict a supersonic flow for $r \gtrsim r_c$. Parker labeled this continuous, supersonic expansion of the corona the "solar wind".

The assumption of an isothermal corona, or $\alpha = 1$ in equation (1.10), is a good approximation only near the sun, where the high coronal temperature is maintained by some (unknown) heating mechanism (III.7). It is certainly a very poor approximation at large heliocentric distances. Parker [1.10] demonstrated the existence of the solar wind class of solutions to the equations of motion (or a steady hydrodynamic expansion of the corona into a vacuum) for the following more realistic assumptions regarding the temperature distribution in the corona:

(1) An isothermal corona in the range $r_0 \leq r \leq b$, and an adiabatic expansion ($\alpha = 5/3$ in equation 1.10) for $r > b$, where b is a heliocentric position beyond the critical radius r_c.

(2) A polytropic corona with

$$\alpha < \tfrac{3}{2}$$

and

$$\alpha < \frac{G M_\odot m}{4 k T_0 r_0} < \frac{\alpha}{2(\alpha - 1)}$$

where $T_0 = T(r_0)$. For $T \approx 10^6\,°\mathrm{K}$ at $r_0 = 10^{11}$, $G M_\odot m / 4 k T_0 r_0 \approx 4$ and only the first condition is meaningful. The observed coronal density variation with position suggests $\alpha \approx 1.1$, so that one would expect the expanding model to apply. If α were taken to be 1.1, the solar wind solutions would occur for all temperatures between 0.73×10^6 and $3.64 \times 10^6\,°\mathrm{K}$.

Thus our use of the isothermal corona as an example has not led to any crucial oversimplification of the problem. The solutions obtained for the more realistic assumptions (1) and (2) above have the desirable property that $u(r)$ approaches a finite limiting expansion speed as $r \to \infty$ (thus avoiding the $u \to \infty$ of the isothermal case).

I.5 The Extension of the Solar Magnetic Field into Interplanetary Space

Parker [1.9] also considered the implication of a continuous coronal expansion with regard to the nature and configuration of the interplanetary magnetic field. Hot coronal plasma would be expected to have an extremely high electrical (as well as thermal) conductivity. In such a fluid, the concept of "frozen-in" magnetic field lines (i.e., very slow diffusion of plasma transverse to the magnetic field), is applicable. The continuous flow of coronal material into interplanetary space then must result in a transport of the solar magnetic field into the interplanetary region. If the sun did not rotate, the resulting magnetic configuration would be extremely simple; a radial coronal expansion as considered above (with neglect of any magnetic forces) would produce magnetic field lines extending radially outward from the sun.

Of course, the sun does rotate with a (latitude dependent) period of ~25 days for a stationary interplanetary observer (or ~27 days for an observer with the orbital motion of the earth). A solar magnetic field line from a given area element on the sun will then be drawn out along the path followed by the fluid parcels emanating from that area. Again consider a coronal expansion that is purely radial in a stationary frame of reference. In a spherical coordinate system (r, ϕ, θ) *rotating* with the sun ($\theta = 0$ along the axis of solar rotation), the solar wind velocity components are

$$U_r = u, \quad \text{the expansion speed,}$$

$$U_\phi = -\omega r \sin\theta,$$

$$U_\theta = 0,$$

where $\omega = 2.7 \times 10^{-6}$ radians sec^{-1} is the angular velocity of solar rotation. The nonradial velocity component U_ϕ is entirely due to transformation to this rotating frame of reference. The path followed by fluid parcels from a given source area is simply a flow streamline (as the source area remains fixed in this frame), determined by the differential equation (see Fig. 1.3).

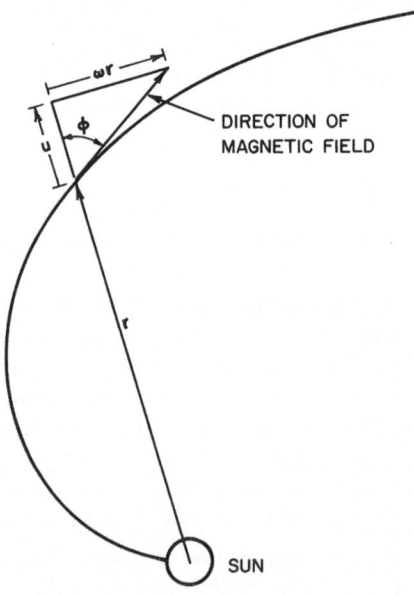

Fig. 1.3   The local orientation of a flow streamline or magnetic field line in a frame of reference rotating with the sun

$$\frac{1}{r}\frac{dr}{d\phi} = \frac{U_r}{U_\phi} = \frac{u}{-\omega r \sin\theta} \tag{1.22}$$

with the condition that θ is constant. The streamlines are also magnetic field lines, so that the latter are defined by (1.22). For heliocentric distances greater than several times r_c, the solar wind solutions of (1.14) predict that $u(r)$ is nearly constant (e. g., Fig. 1.2). In the then reasonable approximation $u(r)=u_s$, a constant, (1.22) is easily integrated to give the explicit form of the magnetic field lines

$$r - r_0 = \frac{-u_s}{\omega \sin\theta}(\phi - \phi_0) \tag{1.23}$$

where ϕ_0 is an initial position at a reference distance r_0. Maxwell's equation $\nabla \cdot \boldsymbol{B} = 0$ for a spherically symmetric geometry leads to the explicit magnetic field components

$$\left.\begin{aligned} B_r(r,\phi,\theta) &= B(r_0,\phi_0,\theta)\left(\frac{r_0}{r}\right)^2, \\[2mm] B_\phi(r,\phi,\theta) &= -B(r_0,\phi_0,\theta)\frac{\omega r_0}{u_s}\frac{r_0}{r}\sin\theta, \\[2mm] B_\theta &= 0. \end{aligned}\right\} \tag{1.24}$$

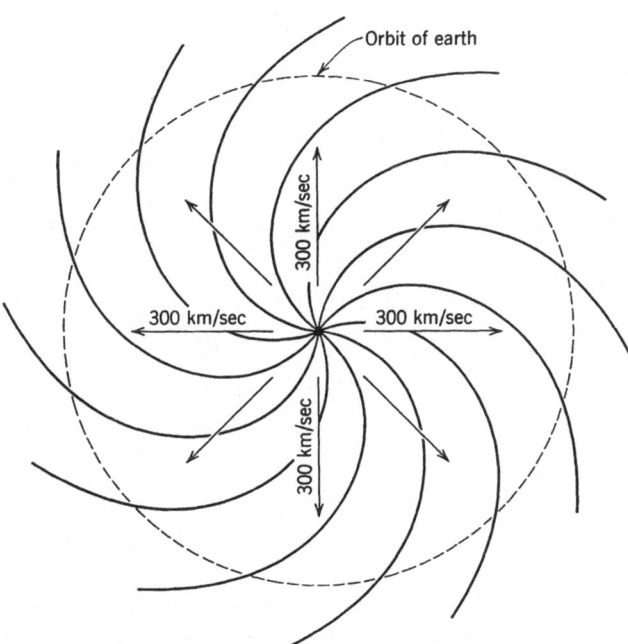

Fig. 1.4 The interplanetary magnetic field configuration in the solar equatorial plane for a constant solar wind speed [1.10]

Fig. 1.4 illustrates this field configuration near the solar equatorial plane. The magnetic field lines are simply drawn into spirals by solar rotation. The sense and strength of the field are determined entirely by the field configuration on the spherical surface at $r=r_0$. Transformation to a stationary frame of reference gives the same field configuration, along with an electric field

$$E = -u \times B = -u_s B_\phi i_\theta$$

where i_θ is a unit vector in the θ direction. This field arises in the stationary frame because the plasma flow is no longer along the spiral field lines.

I.6 An Alternative Model of the Coronal Expansion

The concept of a continuous, supersonic coronal expansion did not win immediate and universal acceptance despite apparent agreement with Biermann's interpretation of comet tail observations. While the supersonic solution for the isothermal corona described in I.4 is unique in satisfying the boundary conditions of small expansion speeds in the corona and vanishing pressure at large heliocentric distances, this uniqueness does not hold for more general temperature distributions. For example, polytropic indices greater than 1 (the isothermal case) in (1.10) lead to solutions similar to those of Class 1 and Class 2. However, the family of Class 2 solutions with $u \to 0$ as $r \to \infty$ also gives $P \to 0$ as $r \to \infty$. Thus no choice between the supersonic, solar wind solutions and these subsonic solutions can be made on the general basis of boundary conditions.

Chamberlain [1.11] considered solutions of the fluid equations (1.7) to (1.9) with the source term due to heat conduction included in the energy equation to give

$$\frac{1}{r^2} \frac{d}{dr}\left[r^2 \rho u \left(\frac{1}{2} u^2 + \frac{3}{2} \frac{P}{\phi} \right) \right] = -\frac{1}{r^2}\frac{d}{dr}(r^2 Pu) - \rho u \frac{GM_\odot}{r^2}$$
$$+ \frac{1}{r^2} \frac{d}{dr}\left(r^2 \kappa_0 T_e^{\frac{5}{2}} \frac{dT_e}{dr} \right). \qquad (1.25)$$

This energy equation has the first integral

$$4\pi \left\{ \rho u r^2 \left[\frac{1}{2} u^2 + \frac{5}{2} \frac{P}{\rho} - \frac{GM_\odot}{r} \right] - r^2 \kappa_0 T_e^{\frac{5}{2}} \frac{dT_e}{dr} \right\} = F, \qquad (1.26)$$

merely stating that the total energy flux through any sun-centered spherical surface has the constant value F. Chamberlain argued that this energy flux, essentially equal to the kinetic energy in the flow of gas at $r \to \infty$, should be zero, and derived numerical solutions under this choice of the integration constant F. These solutions correspond to the Class 1 solutions of I.4, predicting a subsonic expansion at speeds of $\sim 20\,\mathrm{km\,sec}^{-1}$ at $1\,\mathrm{AU}$, with $u \to 0$ and $P \to 0$ as $r \to \infty$. The supersonic expansion solutions of the same equations would result from $F > 0$. The difference between the supersonic and subsonic models was thus reduced to the arbitrary choice of an integration constant.

As a criterion for choosing between the subsonic and supersonic solutions, Chamberlain proposed that the applicable fluid model should exhibit properties similar to those of evaporative models of the corona. Such models follow the motions of *individual* coronal particles under the influence of solar gravity, coulomb collisions, and an electric field produced by polarization of the plasma. The existence of the latter is most easily seen by assuming no interaction between the electrons and protons in the corona and writing the separate equations for hydrostatic equilibrium of the two constituents;

$$\frac{dP_e}{dr} = -\frac{GM_\odot m_e n_e}{r^2} \tag{1.27}$$

and

$$\frac{dP_p}{dr} = -\frac{GM_\odot m_p n_p}{r^2}, \tag{1.28}$$

where $P_e = n_e k T_e$ and $P_p = n_p k T_p$ are the partial pressures of the electrons and protons. Assumption of an isothermal corona (purely for simplicity) would lead to the solutions,

$$n_e(r) = n_{e0} \exp\left\{ -\frac{GM_\odot m_e}{k T_e} \frac{r - r_0}{r r_0} \right\}$$

and

$$n_p(r) = n_{p0} \exp\left\{ -\frac{GM_\odot m_p}{k T_p} \frac{r - r_0}{r r_0} \right\}.$$

Even for equal temperatures, the electron and proton scale heights differ by the ratio m_p/m_e; thus the densities would deviate from equality over any distance comparable to the proton scale height. This would violate the requirement of charge neutrality over any distance greater than the Debye length in a plasma. In fact, any such charge separation would lead to an electric field, changing the hydrostatic equilibrium equations (1.27) and (1.28) to

$$\frac{dP_e}{dr} = -\frac{GM_\odot m_e n_e}{r^2} - n_e q E \tag{1.29}$$

and

$$\frac{dP_p}{dr} = -\frac{GM_\odot m_p n_p}{r^2} + n_p q E, \tag{1.30}$$

where q is the magnitude of the electronic charge. The requirement that $n_e \approx n_p$ and assumption of equal, constant proton and electron temperatures permits determination of the field by subtraction of (1.29) and (1.30):

$$E = \frac{m_p - m_e}{2q} \frac{GM_\odot}{r^2} \approx \frac{m}{2q} \frac{GM_\odot}{r^2}$$

Substitution of this field in (1.29) and (1.30) leads to nearly equal densities for protons and electrons by effectively canceling half of the gravitational force on the heavier protons and adding an equal inward electrostatic force on the lighter electrons. This electrostatic field produced by polarization of a static plasma in a gravitational field is generally known as the Pannekoek-Rosseland Field [1.12, 1.13].

Chamberlain's evaporative coronal model [1.14] used a mean collision time to determine a "critical level" above which coronal ions would move outward from the sun with no further collisions. Such a level exists (even in the *fluid* expansion models described above!) because of the rapid decrease in density with heliocentric distance. In this solar "exosphere", the motions of individual ions in the gravitational and electric fields were followed and mean properties deduced as a function of heliocentric distance. The mean speed of the ions at 1 AU was found to be $\sim 10 \, \text{km sec}^{-1}$. The subsonic solutions of the fluid expansion equations come far closer than the supersonic solutions to matching this evaporative model. Thus, argued Chamberlain, the coronal expansion was more likely to be a moderate, subsonic, "solar breeze," than the supersonic "solar wind" advocated by Parker.

I.7 Confirmation of the Existence of the Solar Wind by *In Situ* Spacecraft Observations

The questions regarding the applicability of supersonic or subsonic expansion models to the solar corona developed into a controversy of some renown, and were by no means resolved before spacecraft technology became sufficiently advanced to send probes into interplanetary space and directly observe its properties. The first *in situ* observations, performed within a few years of 1960, quickly demonstrated that the interplanetary region was pervaded by a supersonic flow of solar plasma, with an average magnetic field configuration similar to that expected in the presence of this flow. Parker's supersonic expansion model clearly gave the better description of the interplanetary plasma and field, and the doubts regarding the model's basic validity and applicability to the solar corona were laid to rest. Curiously, the flaw in Chamberlain's argument for a subsonic solution was not obvious; recent work apparently resolving this dilemma will be described in III.16.

The first *in situ* observations of the solar wind were performed on a series of Russian deep-space probes launched between 1959 and 1961. Plasma collecting cups with a single retarding grid were flown into interplanetary space on the Luna 2, Luna 3, and Venus 3 spacecraft. The flux density of positive ions with energies per charge greater than 50 volts was measured to be several times $10^8 \text{cm}^{-2} \text{sec}^{-1}$ by each spacecraft [1.15, 1.16, 1.17]. No evidence was found for a stationary or slow-moving plasma. The American space probe Explorer 10, launched in 1961, carried a more sophisticated plasma cup capable of measuring the positive ion flux above several energy-per-charge thresholds. A flux density of 1 to $2 \times 10^8 \text{cm}^{-2} \text{sec}^{-1}$ was again observed [1.18]. The extended capability of this instrument led to a determination of a flow speed near 280km sec^{-1} and proton temperatures of 3 to $8 \times 10^5 \, ^\circ\text{K}$ [1.19].

Although these pioneering observations detected a plasma flow similar to that expected in the presence of a solar wind, some possible loopholes and puzzles remained. Both the Russian and American observations covered short periods of time and could not prove beyond doubt that the observed plasma flow was continuous. The Explorer 10 instrument, in fact, observed a plasma flow only intermittently. During the time intervals when no plasma was detected, an anomalously strong magnetic field was observed by the on-board magnetometer. It is now realized that the presence of the geomagnetic field produces a standing bow shock wave in the solar wind (at ~ 15 earth radii along the earth-sun line) and a region of altered flow between the shock front and the boundary of the geomagnetic cavity. By coincidence the orbit of Explorer 10 passed along the average location of the boundary between the geomagnetic field and the region of altered solar wind flow. In retrospect, it appears that Explorer 10 alternately observed the distant geomagnetic field and the plasma flow in the region perturbed by the influence of the earth.

A remarkably successful observational program performed on the Mariner 2 spacecraft, sent to Venus in late 1962, removed any possible remaining doubts as to the existence of the solar wind. Fig. 1.5 shows three-hour averages of the two basic parameters describing the plasma flow, the proton density and flow speed, determined by an electrostatic analyzer system on this spacecraft [1.20, 1.21]. During three months of nearly continuous observation, a plasma flow was always present. The observed proton densities ranged from 0.44 to 54cm^{-3} with an average value $\langle n \rangle = 5.4 \text{cm}^{-3}$. The observed flow speeds ranged from 319 to 771 km sec^{-1}, with an average value $\langle u \rangle = 504 \text{km sec}^{-1}$. The Alfvén Mach number was always found to be greater than two [1.22]. The continuous,

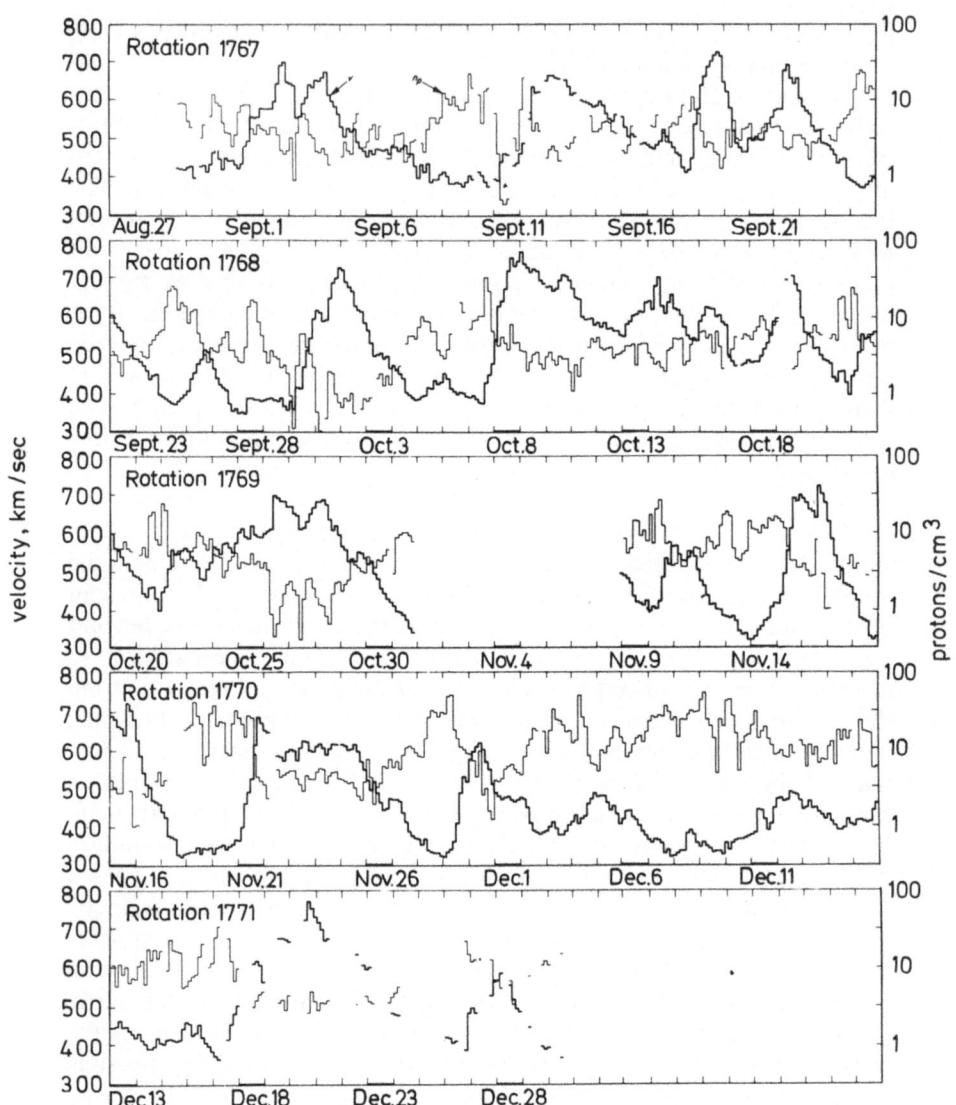

Fig. 1.5 Three-hour averages of the solar wind proton density and flow speed observed by Mariner 2 in 1962 [1.21]. The time coordinate has been broken into 27-day solar rotation periods

supersonic plasma flow observed by Mariner 2 was a striking confirmation of the predictions of Parker's solar wind model.

The configuration of the interplanetary magnetic field was clearly delineated by the magnetometers carried on Mariner 2 and on the Imp 1 satellite, launched into earth orbit in late 1963. Fig. 1.6 shows histo-

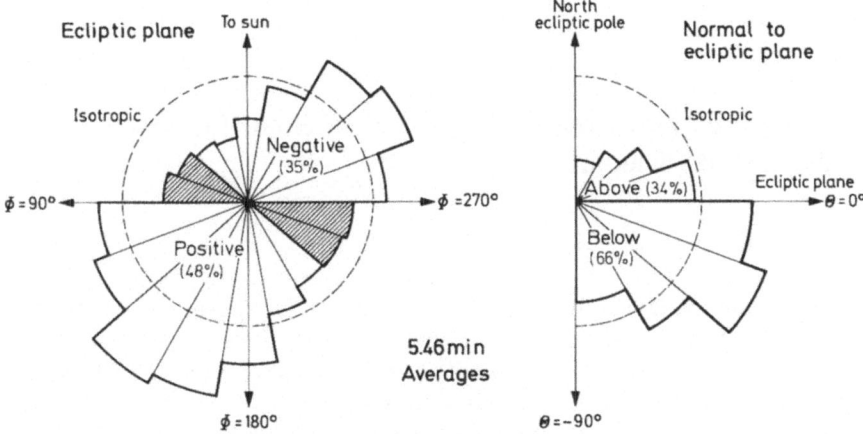

Fig. 1.6   Histograms of the interplanetary magnetic field orientation observed on Imp 1 in 1963 [1.23]. The angle ϕ is the solar ecliptic longitude of the vector field (with $\phi = 0$ the direction toward the sun) and the angle θ is the solar ecliptic latitude of the vector field

grams describing the orientations of 5.46 min averages of the magnetic field observed by Imp 1 [1.23]. The angle ϕ is the solar ecliptic longitude of the observed vector field, while the angle θ is the solar ecliptic latitude. The average field orientation was nearly (but not exactly) in the ecliptic plane, at $\sim 45°$ from the radial direction ($\phi = 0°$ or $180°$). This is in basic agreement with the spiral field configuration of Fig. 1.4. The average field intensity determined on Imp 1 was 6γ, (where $\gamma = 10^{-5}$ gauss is the unit commonly used to specify the weak interplanetary field strength), a reasonable value on the basis of known solar magnetic fields. These observed field properties again confirmed the predictions of the solar wind model.

Chapter II

The Identification and Classification of Some Important Solar Wind Phenomena

II.1 Introduction

The advent of direct spacecraft observations in interplanetary space marked the end of what might be regarded as the first era of solar wind research. This era saw development of the basic concept of the solar wind as well as a model of its origin in the continuous expansion of the solar corona. Observations of a continuous supersonic flow of plasma and a magnetic field configuration fitting the expected pattern by the first spacecraft to probe interplanetary space demonstrated the fundamental validity of this concept and theory.

The subsequent decade has seen the accumulation of an enormous quantity of solar wind data, acquired by spacecraft-borne instruments of growing sophistication. The interplanetary medium, previously a subject of speculation and controversy, has become the most extensively probed of all astronomical plasmas. In a sense, this decade may represent the second era of solar wind research, in which observational knowledge has grown from confirmation of the existence of an interplanetary plasma flow to a fairly detailed description of the dynamical and chemical state of that plasma in the region near the orbit of the earth. The end of this era can be foreseen in the anticipated flood of new observations made over a wide range of heliocentric distances as the space probes of the 1970's are sent as close to the sun as the planet Mercury and as far from the sun as the boundaries of interplanetary space.

Even a cursory look at the results of solar wind observations, such as the Mariner 2 measurements already displayed in Fig. 1.5, reveals a wealth of variations in physical properties on a wide range of time scales. The presence of these variations shows that the coronal expansion is far more complicated than the steady, structureless flow assumed by Parker in his attempt to understand the *basic* characteristics of the interplanetary medium. Such complications were not unexpected, and, in fact, were directly anticipated in the Preface to Parker's monograph [2.1] of 1963. "We will be surprised if future observation does not indicate a restatement of many of the small scale effects that are assumed, and we will be very surprised indeed if observation does not discover a number of complications that have been entirely unanticipated in the present writing. ... But most real theoretical progress can come only *ad hoc* in so complex a dynamical situation as the solar atmosphere and interplanetary space."

The organization of this vast horde of accumulated observations into a coherent phenomenological description, ultimately leading to precise physical interpretation in terms of quantitative models, has become a major task of solar wind research. The primary goal of the present volume is to describe our progress toward such an organization and interpretation in the light of the past decade of spacecraft exploration. Throughout the discussion, emphasis will be placed on the large-scale features of the solar wind and their relationship to solar phenomena. An unfortunate consequence (largely dictated by limitations of space) of this approach is that many small-scale phenomena, such as plasma waves, discontinuities, and instabilities, of great interest to plasma physicists as well as solar wind specialists, will be discussed only in the context of the overall coronal expansion. This may be another sign of the end of an era; the scope of solar wind research has already become so large that a thorough discussion of all important phenomena is no longer feasible in a monograph of moderate length.

II.2 The Classification of Solar Wind Phenomena

A logical first step in the organization of solar wind observations (and in the implementation of our stated intention to concentrate on large-scale features) is the formulation of a scheme for classifying observed phenomena. Such a classification has been proposed by Burlaga and Ness [2.2, 2.3] and applied to magnetic field observations performed near 1 AU. This particular scheme was based upon comparison of the scale time of an observed solar wind phenomenon with three different time scales. The longest of these was taken to be 100 h, essentially the time required for a parcel of 400 km sec^{-1} solar wind to travel from the sun to 1 AU. The other time scales were taken to be 1 h and 0.01 h, or at successive (and largely arbitrary) factors of 10^2 shorter than the first.

We will employ here a classification scheme that is basically similar to that of Burlaga and Ness, but differs in using time scales defined strictly in terms of local physical properties of the interplanetary plasma. Each time scale τ is equivalent to a distance scale λ, where $\lambda = u\tau$ and u is the solar wind speed. In expressing this equivalence, u will be taken as 400 km sec^{-1}, near the mean solar wind speed deduced from most observations.

The longest time scale τ_1 in our classification scheme will be defined as the time required for a parcel of solar wind to flow through a "density scale height." The latter can be defined in analogy to the scale height of an exponential atmosphere (I.3) as

$$\frac{1}{\lambda_1} = \left| \frac{d}{dr} \ln n(r) \right| .$$

The expected near-constancy of the expansion speed in interplanetary space (I.4) and the constancy of particle flux (1.13) imply that $n \propto r^{-2}$, leading to (for $r = r_e$, or 1 AU)

$$\lambda_1 = \frac{r_e}{2} = 0.5 \, \text{AU} = 7.5 \times 10^{12} \, \text{cm}$$

and

$$\tau_1 = \frac{\lambda_1}{u} = 50 \, \text{h} .$$

This definition and value differ from those of Burlaga and Ness by a factor of two resulting from the r^{-2} variation in the density. The density was chosen in defining τ_1 because its dependence upon heliocentric distance can be more plausibly predicted than that of other spatially dependent solar wind parameters.

The intermediate scale time τ_2 in our scheme will be defined in terms of the distance λ_2 over which a sound wave would propagate (in a frame of reference moving with the plasma) during the time required for the plasma to move through the primary scale size, i.e., during the time τ_1. Then

$$\lambda_2 = c_s \tau_1 = \frac{c_s}{u} \lambda_1$$

where c_s is the sound speed, and

$$\tau_2 = \frac{c_s}{u} \tau_1 .$$

Solar wind observations [2.4] show that the average value of c_s/u is close to 1/10, so that

$$\lambda_2 = 0.05 \, \text{AU} = 7.5 \times 10^{11} \, \text{cm}$$

and

$$\tau_2 = 5 \, \text{h} .$$

A similar definition could be given using an Alfvén wave speed, but as the average sound and Alfvén speeds observed at 1 AU are nearly equal, this would lead to a similar value of τ_2.

The shortest scale time τ_3 in our classification scheme will be defined in terms of the proton cyclotron radius at 1 AU. Using $\langle B \rangle = 5 \gamma = 5 \times 10^{-5}$ gauss and a mean proton thermal speed of $40\,\mathrm{km\,sec}^{-1}$ (approximately equal to c_s) gives

$$\lambda_3 = 8 \times 10^6\,\mathrm{cm} \approx 5 \times 10^{-7}\,\mathrm{AU}\,.$$

The corresponding scale time is then

$$\tau_3 = \frac{\lambda_3}{u} = 0.2\,\mathrm{sec}\,.$$

The classification of any observed solar wind phenomenon simply involves comparison of its characteristic scale time, τ, (as recorded by a stationary observer) with the three physical time scales defined above. The physical meaning of the classification follows directly from the physical bases for our definitions of τ_1, τ_2, and τ_3. Seven possible classes of phenomena exist within this framework; these classes are summarized in Table 2.1 and described below.

Table 2.1   *A Classification of Solar Wind Phenomena on the Basis of Characteristic Time Scales*

Time	Physical Scale Time	Class	Type of Phenomena
		1	Steady, Structureless Coronal Expansion
$\tau_1 = 50\,\mathrm{h}$	Expansion	2	Nonsteady, Structured Coronal Expansion
		3	Propagating Structures With Large-Scale Pressure Gradients
$\tau_2 = 5\,\mathrm{h}$	Sonic Motion	4	Propagating Structures Approaching Large-Scale Pressure Equilibrium
		5	Convected Structures With Large-Scale Pressure Equilibrium
$\tau_3 = 5 \times 10^{-4}\,\mathrm{h}$ or 0.2 sec	Proton Cyclotron	6	Structures at Limit of Hydromagnetic Theory
		7	Plasma Waves and "Noise"

Class 1, $\tau \gg \tau_1$: Phenomena involving variations on a time scale much longer than that for the basic coronal expansion. These phenomena can be adequately described in terms of steady, spherically-symmetric models of the coronal expansion. An obvious example would be a completely steady solar wind at 1 AU $(\tau = \infty)$. This condition could arise only if there were no temporal changes or spatial structure in the coronal expansion. Very slow temporal changes in the corona would also be expected to produce phenomena of this class, in an approximation similar to the "adiabatic" approximation in particle [2.5] or quantum mechanics.

Class 2, $\tau \approx \tau_1$: Phenomena involving variations on the same time scale as that for the basic coronal expansion. These phenomena can be adequately described only in terms of nonsteady or spatially structured models of coronal expansion and solar wind. Large scale coronal structures would be expected to produce solar wind phenomena in this class.

Class 3, $\tau_1 \gg \tau \gg \tau_2$: Phenomena involving variations on a time scale much shorter than that for the basic coronal expansion, but with a distance scale longer than that over which a sound signal can propagate in the basic expansion. These phenomena would preserve pressure gradients imposed near the sun, and should be described in terms of propagating disturbances of the basic coronal expansion. Temporal changes in the corona on the time scale of hours would be expected to produce solar wind phenomena in this class.

Class 4, $\tau \approx \tau_2$: Phenomena involving variations on a distance scale comparable to that over which a sound signal can propagate in the basic coronal expansion. Any pressure gradients imposed near the sun would be decaying away, so that such phenomena should be described as propagating, but decaying to a state with no pressure perturbation of the basic expansion.

Class 5, $\tau_2 \gg \tau \gg \tau_3$: Phenomena involving variations on a distance scale much shorter than that over which a sound wave can propagate in the basic expansion, but on a time scale much longer than the cyclotron period. These phenomena (and those in the following classes) would show only local pressure variations that must average out on a large scale. As such, they can propagate only at about the sound speed relative to the plasma, and are largely "convected" through interplanetary space by the supersonic motion of the solar wind. Theoretical treatment as perturbations of a slowly varying background (in a frame of reference moving with the plasma) would be appropriate. Small scale or short period coronal variations, or local, interplanetary processes, could produce phenomena in this class.

Class 6, $\tau \approx \tau_3$: Phenomena involving variations on a time scale comparable to the cyclotron period. Hydromagnetic concepts, valid for all of the above classes, would begin to break down on this time scale. Microscopic, local plasma processes would produce such phenomena.

Class 7, $\tau \ll \tau_3$: Phenomena involving variations on a time scale much less than the cyclotron period. These phenomena belong in the realm of plasma kinetic theory.

II.3 An Identification, Description, and Classification of Some Solar Wind Phenomena

A number of distinct physical phenomena have been identified among the multitude of variations in observed solar wind properties. These phenomena will now be surveyed in the order of descending time scales and related to our proposed classification scheme. If the time scale, τ, is defined in the frame of reference of a stationary (or nearly so) observer, no interpretative prejudices need be imposed in application of this scheme. However, the reader should be forewarned that the rapid flow of the solar wind transports even purely spatial structures (independent of time in a frame of reference moving with the plasma or rotating with the sun) past a stationary observer, producing an apparent temporal variation. A single observer, in fact, cannot distinguish between spatial and temporal effects. The time scale of an observed variation must be understood in this light; it may be equivalent to a distance scale (as in II.2) for a purely or predominantly spatial variation. This effect is extreme for waves propagating through the plasma. Both the apparent period and the propagation direction can be severely distorted. Waves moving toward the sun relative to the plasma may be observed to move outward from the sun by a stationary observer!

Persistent, High-Speed Solar Wind Streams

The variations in solar wind properties observed by Mariner 2 (Fig. 1.5) were dominated by a number of "high-speed streams" [2.6]. The structure of a typical stream (for example, that occurring between Oct. 7 and Oct. 15 on Fig. 1.5) showed a rapid rise in the flow speed, followed by a slower decline. A short period of very high densities near the leading edge of the stream, followed by a longer period of low densities was also evident, as was a variation in proton temperature with a pattern similar to that of the flow speed. The observed magnetic polarity (i.e., the sign of the radial field component) was generally constant within each high-speed stream [2.7]. The duration of a typical stream was five days; a characteristic time for the associated physical variations would be about one-half of this duration, or $\tau \approx 60\,\mathrm{h}$. Thus $\tau \approx \tau_1$, and high-speed streams fit into Class 2 of our classification scheme. They should be interpreted or modeled as part of a structural coronal expansion.

An important clue to the nature of these streams was their apparent tendency to recur at intervals of ~ 27 days. This tendency is emphasized in Fig. 1.5 by the division of the time scale into 27-day intervals, which are displayed one above the other as a vertical column. Features that recur with a 27-day period then fall along a vertical line. The high speed streams of Fig. 1.5 display such a tendency, suggesting that they be interpreted as long-lived *spatial* structures rotating with the sun. If such a persistent stream of high-speed plasma were emitted from a fixed, local coronal source, solar rotation would sweep this stream past an interplanetary observer (moving with the approximate orbital speed of the earth) in each successive rotation period [2.6], as observed.

This interpretation and the observation that geomagnetic activity was high when the Mariner 2 high-speed streams enveloped the earth links the streams with a mystery inherited from pre-spacecraft era studies of solar-terrestrial relationships. Some geomagnetic activity had been found to recur with the solar rotation period and was commonly attributed to particle streams emitted from long-lived solar source regions. Early attempts to identify these hypothetical sources were unsuccessful, and they came to be designated as "*M*-regions" [2.8]. The Mariner 2 observations pointed to recurrent high-speed streams as the interplanetary agent producing the geomagnetic activity, but attempts to identify the *M*-regions producing the streams with known solar phenomena remained unsuccessful [2.9]. The problem of identifying the solar sources of high-speed streams will receive considerable attention in Chapter V.

Flare-Produced Interplanetary Shock Waves

A second type of variation in solar wind properties, also involving a large elevation of the flow speed, is characterized by initial, abrupt rises in the density, flow speed, and proton temperature. An example of such a solar wind disturbance is shown in Fig. 2.1, where the abrupt changes in solar wind properties occurred at ~ 0615 UT on Dec. 18, and were followed by further variations lasting for about 2 days. Densities were high during the first ~ 12 h of the disturbance, sinking to unusually low values thereafter. The proton temperature varied in much the same manner as the flow speed. Fig. 2.2 shows magnetic field observations from Dec. 18; the magnetic field intensity reached unusually high values during the early stages of the disturbance.

The abrupt initial changes in plasma and magnetic properties have been shown to be consistent with the passage of a shock front, at which the kinetic energy of the pre-shock plasma is partially converted (irreversibly) to thermal energy [2.10]. The entire disturbance following such shocks lasts for about two days, leading to a characteristic time scale of about half of this duration, or $\tau \approx 25$ h. Thus $\tau_1 \gtrsim \tau \gg \tau_2$, and these phenomena fit into class 2 or 3 of our classification scheme. This phenomenon had been hypothesized and widely discussed on the basis of the sudden commencement and subsequent development of

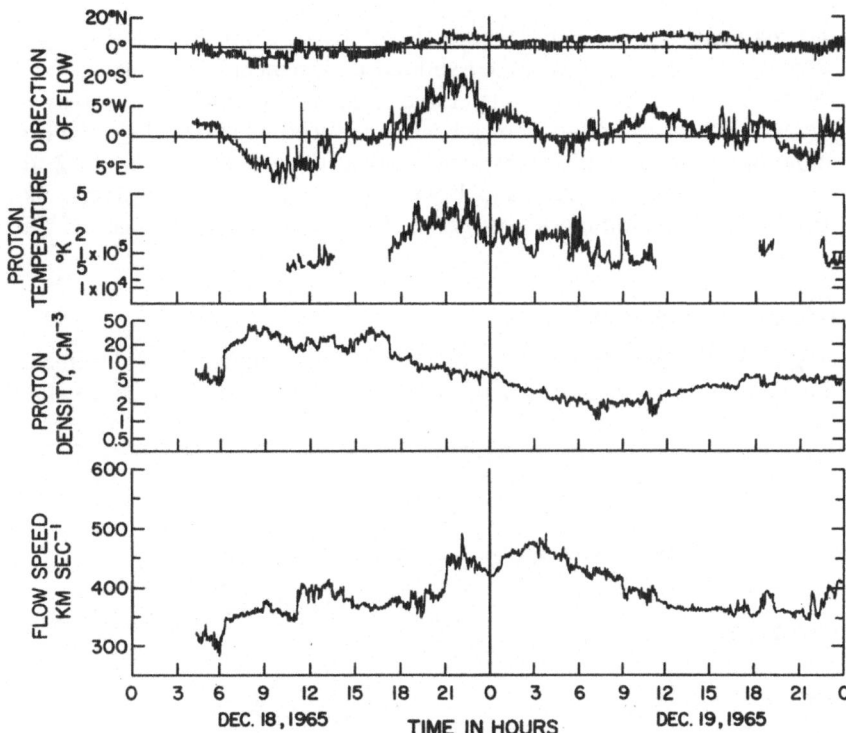

Fig. 2.1 The density, flow velocity, and proton temperature observed by Pioneer 6 on Dec. 18—19, 1965 (unpublished data, courtesy of A. J. Lazarus)

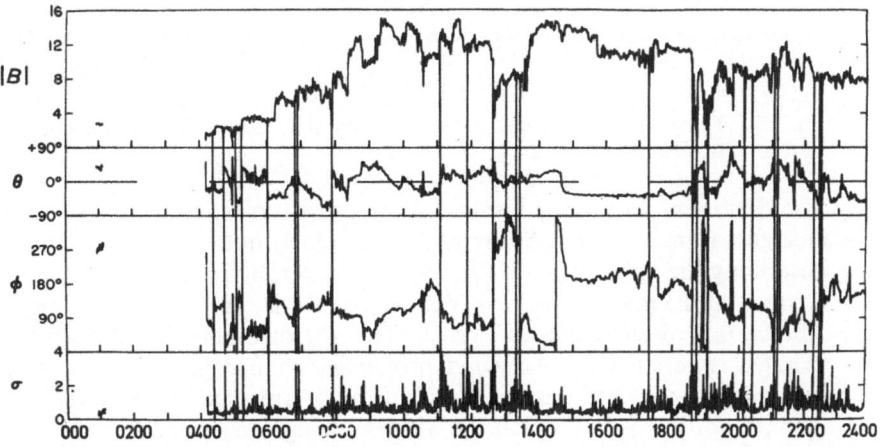

Fig. 2.2 Thirty-second averages of the magnetic field characteristics observed by Pioneer 6 on Dec. 18, 1965 [2.2]. The quantity σ is the standard deviation of the observed fields for each averaging interval

geomagnetic storms long before direct solar wind observations were performed. Many geomagnetic sudden commencements were found to occur one to several days after large solar flares, and it was widely accepted that the entire disturbance was produced by material ejected from solar flares. Solar wind observations such as those in Figs. 2.1 and 2.2 confirm the existence of such shock waves, and, although the precise flare origin is often ambiguous, suggest the long-postulated effect of flare activity.

Interplanetary Filaments

Fig. 2.3 displays 30-second averages of the magnetic field intensity and orientation observed by the Pioneer 6 spacecraft on Dec. 19, 1965.

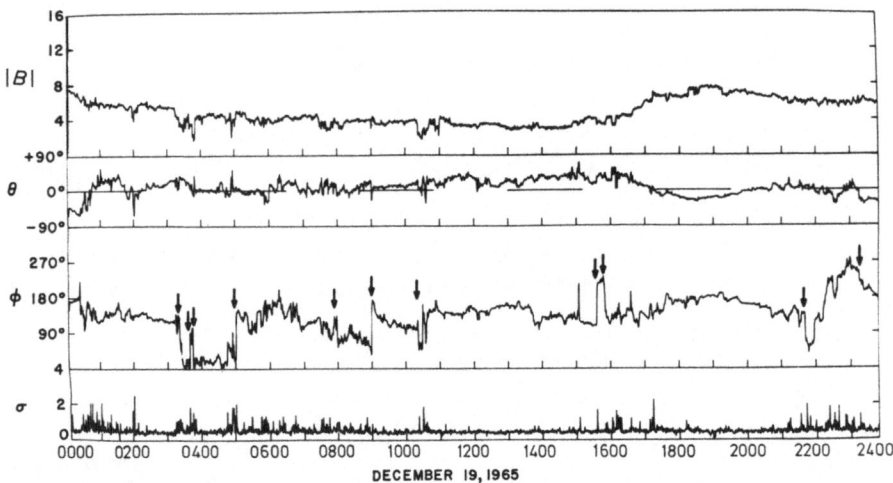

Fig. 2.3   Thirty-second averages of the magnetic field characteristics observed by Pioneer 6 on Dec. 19, 1965 [2.2]

The dominant variations visible in this presentation are discontinuous changes (as indicated by the arrows), particularly noticeable in the angles describing the orientation of the field, but generally not accompanied by changes in the plasma velocity. Variations of this type have been termed "directional discontinuities" [2.2]. The regions between the abrupt changes, wherein the solar wind properties change more slowly, have sometimes been referred to as "filaments," presumably separated by distinct boundaries at which the abrupt changes occur [2.2, 2.11]. The $\sim 1\,\mathrm{h}$ separation of these boundaries in magnetic field data implied a scale size of $\sim 0.01\,\mathrm{AU}$ for the filaments. Thus $\tau_2 \gg \tau \gg \tau_3$, and filaments would fit into Class 5 of our classification. They should then be interpreted as convected solar wind features with large-scale pressure equilibrium on any appreciable distance scale. In fact, Burlaga [2.12] has inferred that the interplanetary regions containing such directional discontinuities are at constant pressure. The filamented character of the interplanetary plasma has been reported to extend to other solar wind properties (perhaps even to the chemical composition). However, the definition of filaments is not precise, and may even depend upon the particular solar wind parameter used as a criterion. For example, plasma properties have been found to show no important changes at a directional discontinuity in the magnetic field [2.13]. A possible interpretation of directional discontinuities as propagating waves (see below) would also deny the long-term identity of the plasma within filaments.

Alfvén Waves

Fig. 2.4 displays both magnetic field and plasma data obtained during a 24-h period by the Mariner 5 spacecraft. The top three frames show five-minute averages of the three components of the plasma velocity and the magnetic field in a cartesian coordinate system with the R coordinate pointing away from the sun, the T coordinate normal to R and parallel to the exliptic plane, and the N coordinate normal to the ecliptic plane. These components are given relative to their average values over the period of display. The magnetic field intensity $|B|$ and the proton density n are shown in the bottom frame of the figure. This display reveals variations in each magnetic field component that are highly correlated with variations

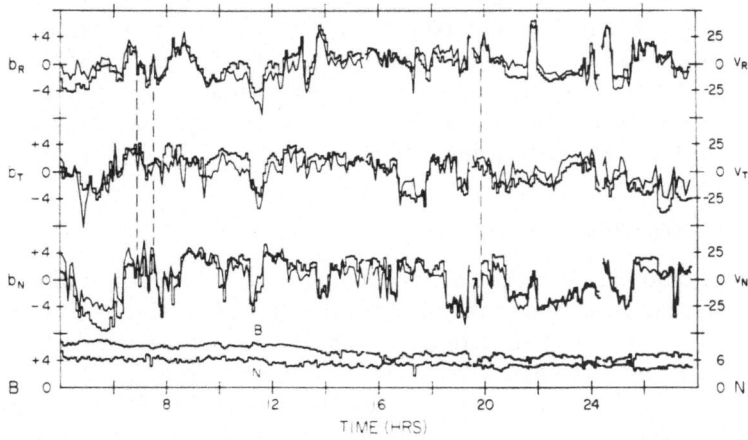

Fig. 2.4   Five-minute averages of the magnetic field and plasma velocity components, magnetic field strength, and proton density observed by Mariner 5 [2.14]

in the same velocity component. Belcher and Davis [2.14] have identified these correlated variations as aperiodic Alfvén waves. The key to this identification is the relationship

$$b = \pm (4\pi\rho)^{\frac{1}{2}} v \qquad (2.1)$$

between the magnetic perturbation b and the velocity perturbation v for an Alfvén wave. The variations in Fig. 2.4 show not only proportionality of b and v but are consistent with the proportionality constant of (2.1) to the accuracy of the observations. This phenomenon is found to be common in the Mariner 5 data. The magnetic perturbations b show a strong statistical tendency to be transverse to the average magnetic field, as would be expected for Alfvén waves. The periods of the observed variations range from ~ 4h down to the limit of temporal resolution, 10 min; this implies wavelengths of 5×10^6 km (or 0.03 AU) and smaller. Thus $\tau_2 \gtrsim \tau \gg \tau_3$, and the observed Alfvén waves fit into Class 5 of our classification scheme. They should then be interpreted as primarily convected features (although they are moving at the Alfvén speed with respect to the plasma) with large scale pressure equilibrium.

The phases between the velocity and magnetic variations indicate that essentially all of the observed waves are moving outward from the sun. If both inward and outward (relative to the plasma) propagating waves were present in the solar wind, both classes would be transported away from the sun by the motion of the plasma, and both should be observed. Unless a mechanism that produces only outward moving waves could be invoked in interplanetary space, this suggests that the waves originated in the sub-Alfvénic region of flow near the sun where the inward moving waves could "escape" into the photosphere. Belcher and Davis suggest that the observed interplanetary Alfvén waves may be the remnants of the waves thought to be responsible for heating of the solar corona (III.7). This interpretation is reasonable as, of the possible wave modes in the corona, Alfvén waves are weakly damped and could travel to 1 AU with little attenuation.

Hydromagnetic Discontinuities

The common occurrence of abrupt changes, or "discontinuities," in solar wind properties has already been mentioned in relation to Fig. 2.3. A small fraction of these changes have been shown to reflect the passage of shock fronts. Two examples of the far more common "directional discontinuity" are shown in parts (a) and (b) of Fig. 2.5. These variations in solar wind properties are not, of course, actually discontinuous. Siscoe *et al.* [2.11] have examined the time required for the observed magnetic changes at a sample of discontinuities, and found that 80% of

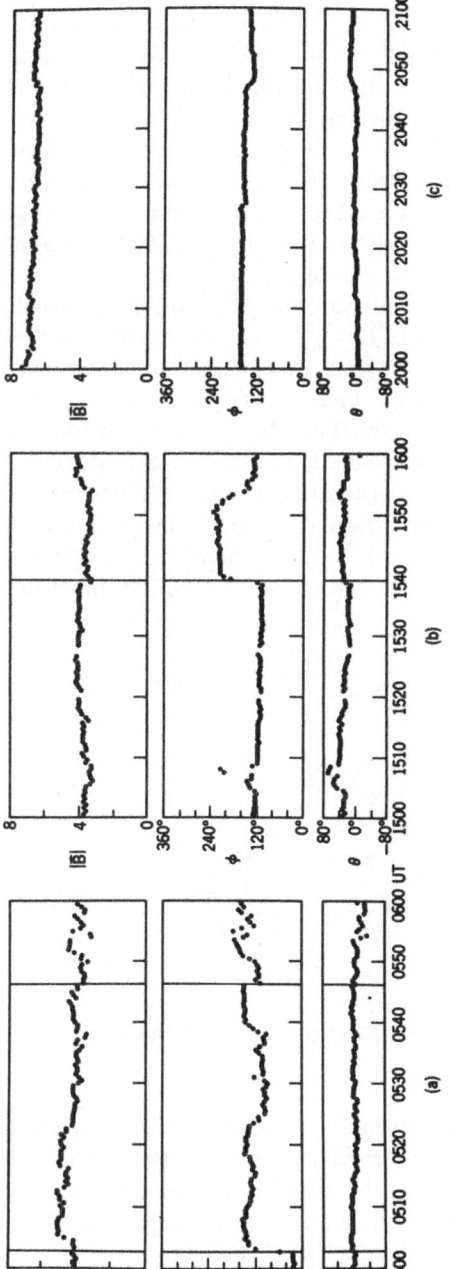

Fig. 2.5 Magnetic field observations from three short time intervals on Fig. 2.3 [2.2]

these times were less than 10 sec. Thus $\tau_2 \gg \tau > \tau_3$, and these discontinuities fit into class 5 of our classification scheme. Hydromagnetic
concepts should be applicable to the relationships between the plasma
states on the two sides of such discontinuities. This time scale corresponds
to a spatial scale size of ~ 4000 km or $\sim 3 \times 10^{-5}$ AU, infinitesimal on
the scale of the overall coronal expansion. In discussing the latter, we
can thus maintain the fiction of a surface at which a discontinuous change
occurs in solar wind properties. This surface can be regarded as planar
in considering the local nature of these changes.

The best known interpretation of directional discontinuities, largely
developed by Burlaga [2.2, 2.3, 2.12, 2.15], is that of surfaces at rest
with respect to the interplanetary plasma, but convected along with
the general motion of the solar wind. These surfaces would then act
as boundaries of distinct plasma regimes. The plasma states on the two
sides of such a boundary would be related, in the context of hydromagnetic theory, by Maxwell's equations and the requirement for pressure
balance normal to the surface. Two types of such *stationary* surfaces
of discontinuity are permitted, but only one of these, the "tangential
discontinuity," has been identified in the solar wind. Fig. 2.6 illustrates

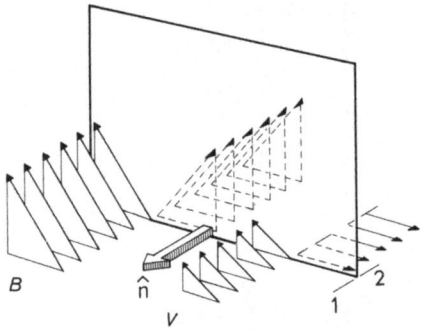

Fig. 2.6   Magnetic field and plasma velocity changes at a tangential discontinuity
with normal $\hat{n}$ [2.16]

the properties of the tangential discontinuity. Its characteristic feature is
that the magnetic field B and the flow velocity v (in the frame of reference
moving with the discontinuity) must be parallel to the surface. The
magnitude and orientation of B and v can change in any manner consistent
with this restriction. In an isotropic, neutral, proton-electron gas,

$$P = nk(T_e + T_p) + \frac{B^2}{8\pi}. \tag{2.2}$$

If $B = |B|$ changes at a tangential discontinuity, n, T_e, and T_p must change to maintain the constancy of P. Burlaga has demonstrated that the plasma and field changes at many observed discontinuities are consistent with these requirements (although the pressure balance has not been tested with explicit knowledge of the electron temperatures).

Another possible type of discontinuity surface, referred to as a "rotational discontinuity," *propagates* through a plasma at the Alfvén speed and can thus be thought of as a sharply crested Alfvén wave. Fig. 2.7 illustrates the characteristics of a rotational discontinuity.

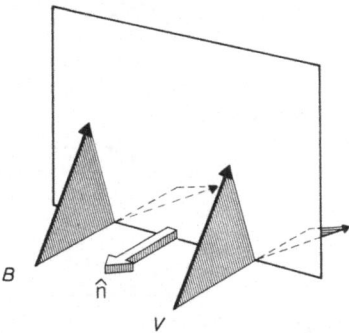

Fig. 2.7   Alfvén shock. Magnetic field and plasma velocity changes at a rotational discontinuity with normal $\hat{n}$ [2.16]

The components of B and v normal to the surface are not zero, so that (in contrast to the tangential discontinuity) both field lines and plasma pass through the discontinuity. In an isotropic plasma, both the magnetic field intensity and the plasma velocity are the same on the two sides of the surface. The changes in these two vectors associated with the passage of a rotational discontinuity are then merely "rotations" in the discontinuity surface (hence the name). These changes are related precisely as in (2.1). Pressure balance (there is motion, but no acceleration across the surface) is expressed as in (2.2). However, if both B and (to maintain a constant particle flux) n remain unchanged, (2.2) simply implies that the sum of electron and proton temperatures is also unchanged. Thus the rotational discontinuity in an isotropic plasma is nothing more than a sharp "kink" propagating along a magnetic field line. This simplicity is modified in an anisotropic plasma, but only slightly for the typical anisotropies of the total pressure that are observed in the solar wind (see III.3). Fig. 2.8 shows an observed abrupt change in solar wind properties that has been interpreted as a rotational discontinuity by Belcher and Davis [2.14]. The change in the observed proton density at the discontinuity is probably smaller than the accuracy of the density determination.

The interpretations of directional discontinuities in the interplanetary magnetic field as either tangential or rotational lead to contrasting concepts of the fine structure of interplanetary space. If these changes

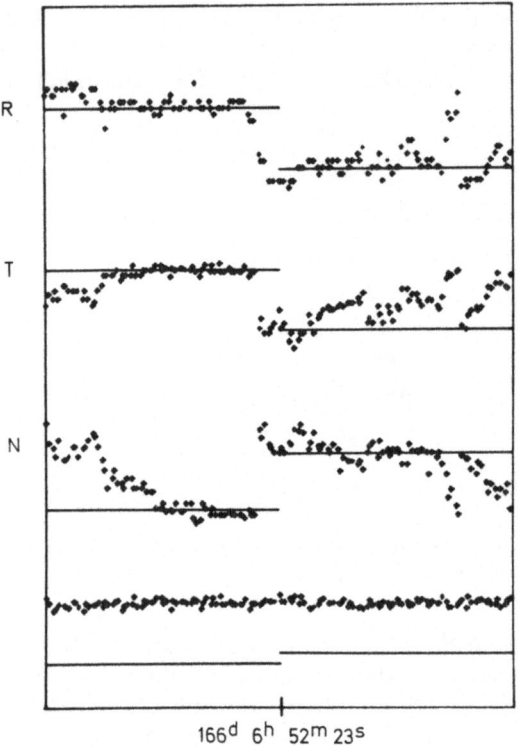

$$166^d \ 6^h \ 52^m \ 23^s$$

Fig. 2.8   A rotational discontinuity identified in Mariner 5 solar wind observations by Belcher and Davis [2.14]. The Cartesian components R, T, and N of the magnetic field (dots) and plasma velocity (horizontal lines) are defined in the text. The lowest frame shows the field magnitude and plasma density

reflect the presence of tangential discontinuities, the interplanetary plasma consists of distinct, bounded plasma regimes (perhaps the "filaments" mentioned above) that retain their identity in the absence of instabilities and diffusion (processes that would be inhibited by the tangential nature of the field lines at boundary surfaces). If these changes reflect the presence of rotational discontinuities, the interplanetary plasma is permeated with propagating magnetic irregularities that do not define distinct plasma regimes. It is clear that both types of discontinuities exist in the solar wind. However, considerable controversy presently surrounds the question of which phenomenon is most common. Our discussion of the large scale features of the coronal expansion will not (fortunately) hinge upon this question.

II.4 Summary and Plan of Action

The five phenomena identified and briefly described in II.3 have given us examples of Classes 2, 3, and 5 of our classification scheme. Class 1, the essentially steady and structureless solar wind, might be expected to emerge when the larger scale variations of Classes 2 and 4 are absent. The identification and description of this important solar wind state will receive special attention in Chapter 3. Examples of Class 4 are absent. Burlaga and Ogilvie [2.17] have found that pressure variations (and thus dynamical processes) occur on times scales >2 days, but not on time scales near an hour. Thus some scale time such as τ_2, at which pressure variations are decaying, probably does exist. The observation of phenomena on this time scale would be of great interest. Classes 6 and 7 are near the limit of time resolution for present day plasma detectors; for a discussion of magnetic and electric field observations in these classes, the reader is referred to a recent review by Scarf [2.18].

Given our stated interest in the large-scale aspects of the coronal expansion, we will be primarily concerned with phenomena from Classes 1 through 4. The three observed small-scale phenomena of Class 5 will not be discussed for their intrinsic interest (however great that may be), but will be invoked when they bear upon the physics of the larger-scale features. In the following four chapters we will discuss four distinct large-scale phenomena: the dynamics of a structureless coronal expansion the chemical composition of the expanding corona, the interplanetary evolution and solar origin of high speed plasma streams, and the propagation of shock waves through the solar wind. Detailed observational descriptions will be combined with theoretical models in formulating our physical interpretations.

Chapter III

The Dynamics of a Structureless Coronal Expansion

III.1 Introduction

The dynamics of a solar wind free of structure, as would result from a steady, spherically-symmetric coronal expansion, merits primary attention in any discussion of the physics of the interplanetary plasma. We have already seen that the models of Parker, based on the assumption of steady, spherically-symmetric fluid motion, correctly predicted the basic characteristics of the solar wind. The adoption of this same simplifying assumption in the formulation of models attempting a more detailed description of the interplanetary plasma reduces the *mathematical* difficulties inherent in the fluid equations to their simplest level, permitting the consideration of subtle *physical* effects. Such models have been extended to include the forces due to viscosity and the magnetic field, as well as the energy sources due to heat conduction and the dissipation of hydromagnetic waves. The detailed comparison of the predictions of these sophisticated coronal expansion models with solar wind observations should in principle allow one to judge the significance of the added forces and energy sources, or to identify any necessity for including additional physical processes.

These extensions of Parker's basic theory and comparison of their predictions with observations have probably been the most actively pursued areas of solar wind research, and have led to the most fundamental discussions of interplanetary physics. In this light it may be somewhat surprising to realize that the application of such models to the actual solar wind is justified only under rare circumstances. We have already concluded that the solar wind should be regarded as structureless only when there are no variations on the basic time (or distance) scale of the coronal expansion, about 50 h (or 0.5 AU). Yet our examination of solar wind observations in Chapter II has revealed that variations in solar wind properties are almost ubiquitous. In attempting to compare models and observations, we must take care to isolate in the latter the structureless state assumed in the former.

Thus the present chapter will begin with a search for the structureless or "quiet" state of the solar wind and a tabulation of the physical properties that appear to be most pertinent to this ideal state. Two basic models of the steady, radial expansion of a spherically-symmetric corona will then be presented in detail. Comparison of these models with coronal and solar wind observations will lead to discussion of more recent and sophisticated modifications of these basic models. Consideration will finally be given to a recent revival of interest in evaporative coronal models.

III.2 In Search of the Structureless Solar Wind

The first attempt to identify a structureless state of the solar wind was made on the basis of the Mariner 2 observations shown in Fig. 1.5. Neugebauer and Snyder [3.1] suggested that "... during this period the interplanetary plasma consisted of a series of long-lived, high velocity streams separated by slower-moving plasma." The latter was selected for comparison with coronal expansion models; it was concluded that "... the quiet, between-stream, solar wind velocity was in the range 320 to ~340 km sec^{-1}." This postulated quiet state of the solar wind was observed during only five intervals, none of which persisted for longer than three days, during the entire $4\frac{1}{2}$ months of Mariner 2 data acquisition.

Later discussions of solar wind observations have generally followed the lead of Neugebauer and Snyder in associating low-speed solar wind with a quiet state. Fig. 3.1 shows three-hour averages of the flow speeds observed by the twin Vela 2 spacecraft during a single 27-day solar rotation period in early 1965 [3.2]. Between April 1 and April 15, the flow speed was nearly always observed to be near 320 km sec^{-1} (note,

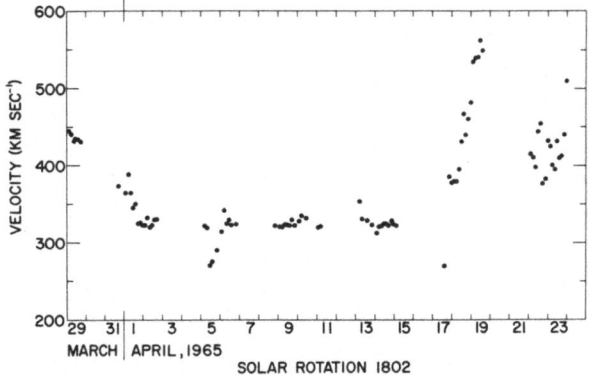

Fig. 3.1 Three-hour averages of the solar wind flow speed observed by the Vela 2 spacecraft during a single 27-day solar rotation in 1965 [3.2]

however, that data gaps do exist). These observations demonstrate that low-speed solar wind does *occasionally* fit the criterion for a steady phenomenon by displaying little change on the expansion time scale of $\sim 50\,\mathrm{h}$.

Flow speeds below $320\,\mathrm{km\,sec^{-1}}$ were found only rarely in the Mariner 2 and Vela 2 observations, leading to the supposition that this formed a "base level" for the coronal expansion [3.3]. However, subsequent observations have clearly shown that the flow speed can sink below $320\,\mathrm{km\,sec^{-1}}$, achieving values as low as $\sim 250\,\mathrm{km\,sec^{-1}}$ on infrequent (and short-lived) occasions. Burlaga and Ogilvie [3.4] have emphasized the absence of any clearly defined base or "quiet" state, suggesting rather that a whole continuum of quiet solar wind states at different, but still low, flow speeds, exists in the interplanetary plasma.

Despite this general agreement among solar wind observers that low solar wind speeds are in some sense indicative of a "quiet state," quantitative evidence supporting this view was not presented until recently. The first such evidence was based on consideration of the difference Δu between consecutive three-hour averages $u(t_0)$ and $u(t_0 + 3\mathrm{h})$ of flow speeds observed by Vela 3 spacecraft [3.5]. Fig. 3.2 shows the root-mean-square difference Δu_{rms} computed for all cases

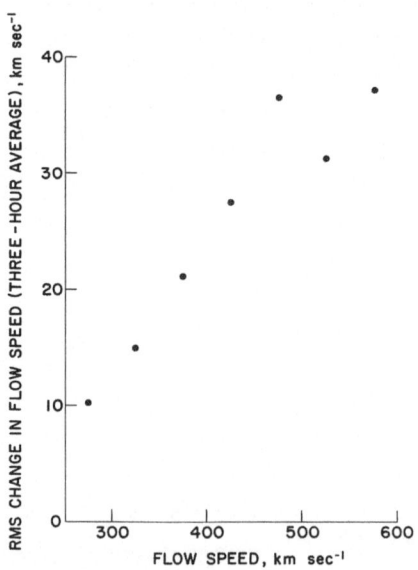

Fig. 3.2 The rms change between consecutive three-hour averages of the flow speed, as a function of (initial) flow speed [3.5]

with $250 \leq u(t_0) < 300 \, \text{km sec}^{-1}$, all cases with $300 \leq u(t_0) < 350 \, \text{km sec}^{-1}$, etc. The parameter Δ_{rms} is then the magnitude of the average variation in flow speed on a three-hour time scale. Fig. 3.2 reveals that such variations are smallest at low flow speeds, and significantly larger at high flow speeds. This result is in accord with the general association of low flow speeds with quiet conditions. However, it still gives us no information regarding variations on the 50 h time scale that is the relevant criterion for a structureless coronal expansion.

The analysis of solar wind speed variations has recently been extended to longer time intervals by Gosling and Bame [3.6] using Vela 2 and Vela 3 observations from 1964 to 1967. This discussion considers pairs of three-hour flow speed averages $u(t_0)$ and $u(t_0 + L)$ where L is an arbitrary time lag (in multiples of the three-hour averaging interval). On Fig. 3.3, $u(t_0)$ is shown as the abscissa and $u(t_0 + L)$ as the ordinate for seven different time lags. If the solar wind varied little on the time scale L, $u(t_0)$ and $u(t_0 + L)$ would be nearly equal and thus lie near a 45° line on the figure. For $L = 3$ h, the first frame of Fig. 3.3, the points do lie near this line, implying that the differences between successive three-hour averages of the solar wind speed are generally small. At larger values of L, the points in Fig. 3.3 scatter away from the 45° line, indicating greater differences between the observed flow speeds. For $L = 72$ h, there is no strong tendency for $u(t_0)$ and $u(t_0 + 72 \text{h})$ to be equal. The variations of u on a 72-hour time scale must be large.

A quantitative measure of the tendency for $u(t_0)$ and $u(t_0 + L)$ to be equal is given by the autocorrelation coefficient. Let us regard $u_1 = u(t_0)$ as an independent variable and $u_2 = u(t_0 + L)$ as a dependent variable. Let σ_M be the standard deviation of the observed u_2 about their mean value $u_M = \langle u_2 \rangle$, and let σ_c be the standard deviation of $u_2 - u_c$, where $u_c(u_1)$ is the linear function that gives the best fit (in the sense of least-squares) to the observed $u_2(u_1)$. Then the autocorrelation coefficient can be defined as [3.7]

$$A_L = \pm \left(\frac{\sigma_M^2 - \sigma_c^2}{\sigma_M^2} \right)^{\frac{1}{2}}.$$

A_L can, in fact, be shown to be the slope of the best fit line in a graph of the points (u_1, u_2), such as Fig. 3.3. If A_L is close to unity, the best fit line is near $u_2 = u_1$, implying that $u(t_0 + L) \approx u(t_0)$. In other words, u has varied only slightly on the time scale L. In contrast, if A_L is equal to zero, the best fit is a horizontal line, $u_2 = u_M$ (as $\sigma_c = \sigma_M$), implying that $u(t_0 + L)$ is, on the average, independent of $u(t_0)$. In other words, u has varied so drastically on the time scale L that no relationship exists between values observed at this temporal separation.

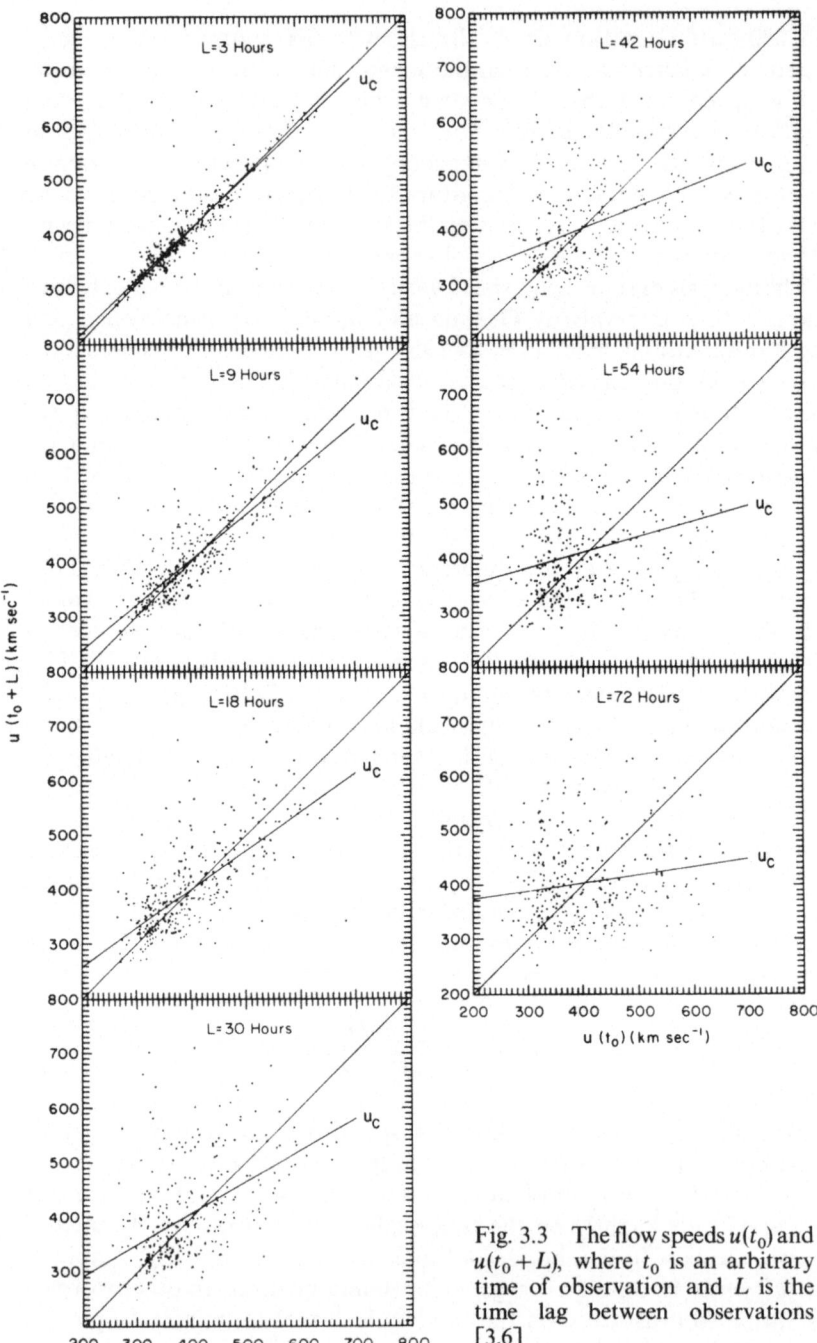

Fig. 3.3   The flow speeds $u(t_0)$ and $u(t_0 + L)$, where t_0 is an arbitrary time of observation and L is the time lag between observations [3.6]

The best fit lines for the different time lags are shown in Fig. 3.3. As would be expected, the increased scatter at longer time lags produces best-fit lines with a decreasing slope, indicating that the autocorrelation coefficient A_L becomes small at long time lags. Fig. 3.4 shows A_L at lags between 0 and 120 h. For the 50-hour time scale of the coronal expansion, the autocorrelation coefficient is only about 0.3. Flow speeds observed with this temporal separation are only slightly related. This demonstrates quantitatively the same conclusion drawn by visual inspection of the data shown in II.3—that the solar wind cannot be regarded as steady under average or typical conditions. Large variations do occur on the basic time scale of the coronal expansion.

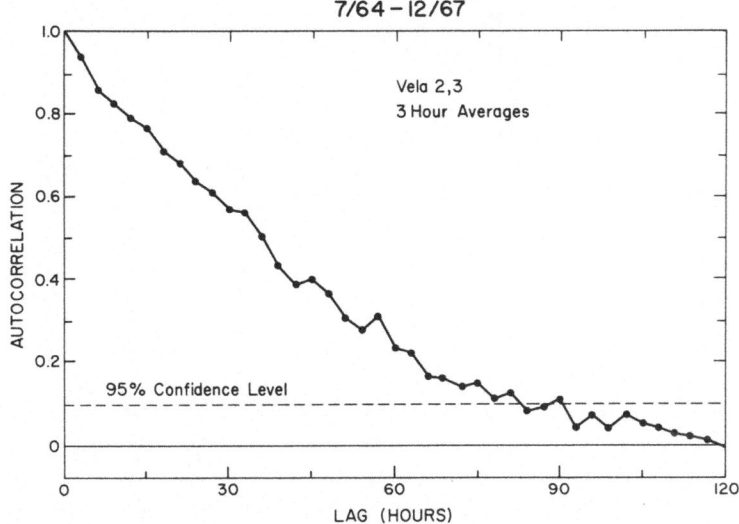

Fig. 3.4   The autocorrelation of flow speed observations separated by time lags between 0 and 120 h [3.6]

A further result of the variation analysis of Gosling and Bame is displayed in Fig. 3.5. Here σ_M^2 is shown for time lags between 0 and 130 h. The deviation from the best-fit line has been computed separately for two classes of observations; σ_{CL}^2 is computed only from "low-speed" solar wind, defined by $u(t_0) < 404 \, \text{km sec}^{-1}$, while σ_{CH}^2 is computed only from "high-speed" solar wind, defined by $u(t_0) > 404 \, \text{km sec}^{-1}$. At short time lags, $\sigma_{CL}^2 < \sigma_{CH}^2$, indicating that $u(t_0)$ and $u(t_0 + L)$ are more nearly equal at low solar wind speeds $u(t_0)$ than at high solar wind speeds. This is in agreement with the result shown in Fig. 3.2. However, at time lags near 50 h, $c_{CL}^2 \approx \sigma_{CH}^2 \approx c_M^2$. The variations on a 50-hour time scale are large even when attention is restricted to low solar wind speeds.

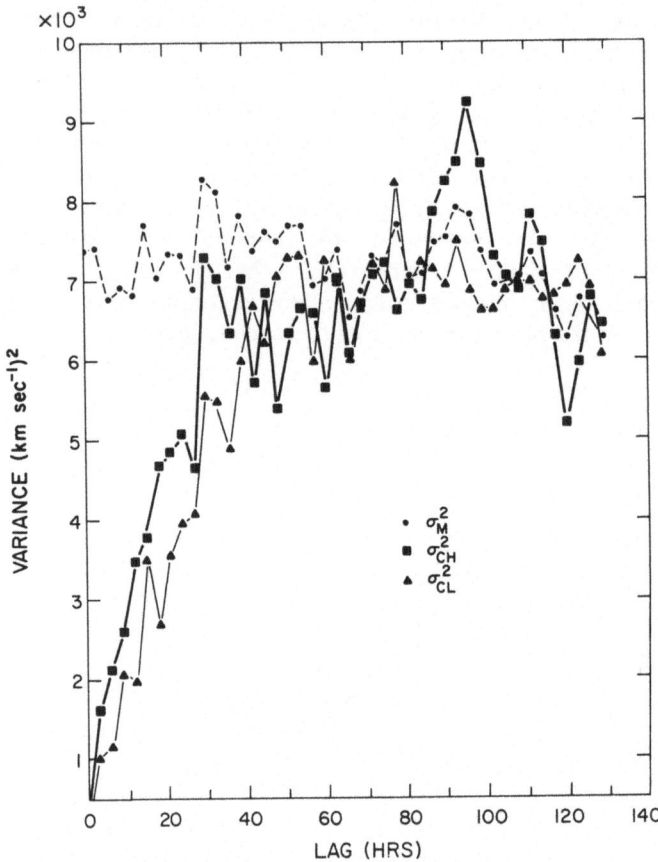

Fig. 3.5   The standard deviations of the best-fit lines to $u(t_0 + L)$ vs. $u(t_0)$: for all observations considered by Gosling and Bame [3.6], σ_M^2; for observations with $u(t_0) < 404\,\mathrm{km\,sec^{-1}}$, σ_{CL}^2; and for observations with $u(t_0) > 404\,\mathrm{km\,sec^{-1}}$, σ_{CH}^2

This analysis lends little support to the supposition that low-speed solar wind corresponds to a structureless coronal expansion.

At this point it is not clear that a structureless state of the solar wind has been identified (in any broad sense) in solar wind observations. In view of the great efforts expended by devoted theorists in deriving highly sophisticated models of a steady, spherically-symmetric coronal expansion, this situation is most unfortunate. It might seem appropriate to examine solar wind observations carefully in an attempt to isolate those occasions on which the solar wind is nearly steady for periods long compared to 50 h. The average solar wind properties computed from such periods might safely be compared with models of the structureless solar wind. In fact, such occasions appear to be sufficiently rare (e.g., see [3.8]) that the statistical validity of averages so computed would be questionable. We are thus forced by necessity to follow here the usual practices of regarding low-speed solar wind as "quiet" and comparing its properties with the predictions of the models. This must remain an unsatisfactory compromise with reality, and should be specifically recognized as such. The resulting comparisons may lead to a valid interpretation of basic physical processes affecting the coronal expansion. They may also be confused by the unidentified effects of the temporal variations and spatial structures that we have seen to be extremely common in the real solar wind.

III.3 The Physical Properties of Low-Speed Solar Wind

Pursuit of our qualified identification of low-speed solar wind as the structureless coronal expansion leads to several practical difficulties. As other solar wind properties are observed to depend upon the flow speed (V.7) the precise choice of a "low-speed" criterion determines the density, proton temperature, etc., with which the theoretical models will be compared. Fig. 3.6 is a histogram of the nearly 14000 individual observations of the flow speed made by the twin Vela 3 spacecraft over a two-year interval. The rapidly decreasing frequency of observations for $u < 350 \, \mathrm{km \, sec^{-1}}$ forces an additional compromise in selecting a low-speed criterion for quiet solar wind conditions. If a flow-speed interval near the lower limit of the range of observations shown in Fig. 3.6 were selected, so few observations would be available as to cast doubt on the statistical significance of the resulting solar wind state. Further, the very rarity of such observations would appear to indicate that solar wind speeds near the lower limit of $\sim 250 \, \mathrm{km \, sec^{-1}}$, when observed, do not

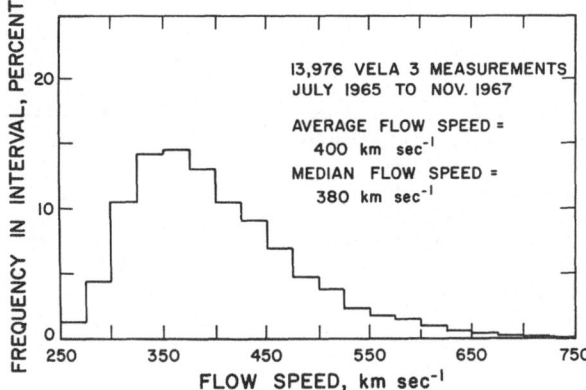

Fig. 3.6   A histogram of the flow speeds observed by Vela 3 spacecraft [3.9]

persist for a sufficiently long time to be indicative of a steady or structureless flow. We will avoid this problem by using the solar wind speed interval 300 to 325 km sec^{-1} as our criterion for low-speed solar wind. This choice includes the "quiet, between-stream" conditions discussed by Neugebauer and Snyder, and the "base-level" found in the Vela 2 observations, and is thus in accord with historical precedents.

Even when attention is confined to solar wind observations with the flow speed in the range 300 to 325 km sec^{-1}, physical parameters such as the density, proton temperature, and magnetic field strength vary widely. For example, the standard deviation (about the average) of the proton densities observed by Vela 3 is as large for the 3 400 cases with $300 \leq u \leq 325$ km sec^{-1} as for all 14 000 observations collected in two years. Restricting attention to low solar wind speeds still does not yield a single, well-defined, and distinct state of the solar wind. The "low-speed" state of the solar wind will thus be defined by using *average values* of solar wind properties under the restriction $300 \leq u < 325$ km sec^{-1}. Table 3.1 displays such values, collected from several sources [3.9, 3.10, 3.11], of a number of important solar wind properties. This tabulation reflects an additional but minor complication—that the random motions of interplanetary particles need not be isotropic. Rather, different temperatures can (and on the average do) apply parallel and transverse to the magnetic field [3.2]. This effect is of only minor importance to the overall dynamics of the solar wind because the total pressure tensor, dominated by the electrons, deviates from isotropy by only $\sim 30\%$ on the average. The random motions can also differ significantly in the two directions along the field lines, i.e., the velocity distribution functions can be "skewed." These characteristics, as well as the nonequality of the average electron and proton temperatures, indicate that the interplanetary plasma is not in a state of thermodynamic equilibrium.

Table 3.1   *Average Properties of the Low-Speed Solar Wind*

Radial Component of Flow Velocity[a]	300—325 km sec^{-1}
Nonradial Component of Flow Velocity[b]	8 km sec^{-1}
Proton (or Electron) Density	8.7 cm^{-3}
Electron Temperature: Average	1.5×10^5 °K
Electron Thermal Anisotropy[c]	1.1
Proton Temperature: Average	4×10^4 °K
Proton Thermal Anisotropy[c]	2
Magnetic Field Intensity	5γ
Solar Ecliptic Longitude of Field	140°

[a] Definition of "Low-Speed".
[b] See III.15.
[c] Ratio of Temperatures Parallel and Transverse to the Magnetic Field.

The energetics of the coronal expansion will be considered both in the present context of the structureless solar wind and in later discussions of high-speed streams and interplanetary shocks. Table 3.2 summarizes the observed particle and energy flux densities relevant to such considerations. It should be emphasized that the heat conduction flux densities given for both electrons and protons are derived by direct integration of the observed fluxes of random energy toward and away from the sun (along field lines). They are purely observational, depending on no transport theory or solar wind model.

Table 3.2 *Average Particle and Energy Flux Densities in the Low Speed Solar Wind*

Proton Flux Density	2.4×10^8 cm^{-2} sec^{-1}
Kinetic Energy Flux Density	0.22 ergs cm^{-2} sec^{-1}
Enthalpy Flux Density	0.008 ergs cm^{-2} sec^{-1}
Gravitational Flux Density	0.004 ergs cm^{-2} sec^{-1}
Magnetic Energy Flux Density	0.003 ergs cm^{-2} sec^{-1}
Electron Heat Conduction Flux Density	0.007 ergs cm^{-2} sec^{-1}
Proton Heat Conduction Flux Density	~ 0.00001 ergs cm^{-2} sec^{-1}

III.4 A "One-Fluid" Model of the Steady, Spherically-Symmetric Coronal Expansion

Consider (as in Chapter I) the steady, radial, spherically symmetric motion of a fully ionized, but electrically neutral, hydrogen plasma in the solar gravitational field. If all magnetic effects are neglected, the fluid equations for mass, momentum, and energy conservation are

$$\frac{1}{r^2} \frac{d}{dr}(\rho u r^2) = 0 , \tag{3.1}$$

$$\rho u \frac{du}{dr} = -\frac{dP}{dr} - \rho \frac{GM_\odot}{r^2} , \tag{3.2}$$

and

$$\frac{1}{r^2} \frac{d}{dr}\left[\rho u r^2 \left(\frac{1}{2} u^2 + \frac{3}{2} \frac{P}{\rho}\right)\right] = -\frac{1}{r^2} \frac{d}{dr}(P u r^2) - \rho u \frac{GM_\odot}{r^2} + S(r) . \tag{3.3}$$

The mass density is given by

$$\rho = (m_p + m_e)n = mn , \tag{3.4}$$

where n is the number density of protons *or* electrons. If the electron and proton temperatures are assumed to be equal, the pressure is

$$P = 2nkT . \tag{3.5}$$

The assumption of equal proton and electron temperatures leads to use of the term "one-fluid" to describe models based on equations (3.1) to

(3.5). If heat conduction is taken to be the only important energy source beyond a heliocentric distance r_0, then

$$S(r) = -\frac{1}{r^2}\frac{d}{dr}(r^2 f_c)$$

for $r > r_0$. The heat conduction flux density f_c (purely radial) is assumed to be related to the temperature gradient by a thermal conductivity κ,

$$f_c = -\kappa\frac{dT}{dr}, \tag{3.6}$$

with

$$\kappa = \kappa_0 T^{\frac{5}{2}}, \tag{3.7}$$

as in I.3. We are thus led to the system of equations

$$\frac{1}{r^2}\frac{d}{dr}(nmur^2) = 0, \tag{3.8}$$

$$nmu\frac{du}{dr} = -2\frac{d}{dr}(nkT) - n\frac{GM_\odot m}{r^2}, \tag{3.9}$$

and

$$\frac{1}{r^2}\frac{d}{dr}\left[nmur^2\left(\frac{1}{2}u^2 + 5\frac{kT}{m}\right)\right] = -nu\frac{GM_\odot m}{r^2} - \frac{1}{r^2}\frac{d}{dr}\left(r^2\kappa_0 T^{\frac{5}{2}}\frac{dT}{dr}\right). \tag{3.10}$$

First integrals of the mass (3.8) and energy (3.10) conservation equations are

$$I = 4\pi nur^2, \tag{3.11}$$

and

$$F = 4\pi nur^2\left[\frac{1}{2}mu^2 + 5kT - \frac{GM_\odot m}{r^2}\right] - 4\pi r^2\kappa_0 T^{\frac{5}{2}}\frac{dT}{dr} \tag{3.12}$$

where the constants I and F are the particle and energy fluxes through a sun-centered sphere of radius r.

Equations (3.9), (3.11), and (3.12) were used by Chamberlain [3.12] in deriving his model of a subsonic coronal expansion (I.6). Whereas Chamberlain restricted attention to a subsonic expansion by requiring $F = 0$ (i.e., $u \to 0$ as $r \to \infty$), we will be interested in the supersonic expansion implied by $F > 0$ (i.e., with u finite as $r \to \infty$). The system of equations (3.9), (3.11), and (3.12) consists of one algebraic equation and two ordinary, first-order, differential equations valid for $r > r_0$, where r_0 is some position in the solar corona. Two integration constants, I and F, are already present, and two more will result from integration of the remaining differential equations. Hence four boundary conditions must be stated to determine a specific solution. Choice of a coronal density $n_0 = n(r_0)$ and temperature $T_0 = T(r_0)$ and the requirement that the solution pass smoothly through the critical point (I.4) with $du/dr > 0$ (i.e., be a "solar wind" solution) provide three of these conditions. The fourth has usually been stated as a restriction on $T(r)$ at large heliocentric distances; this restriction has traditionally been that $T \to 0$ as $r \to \infty$. Our discussion of solar wind and solar breeze solutions in I.4 and I.6 led to the realization that solutions with very similar behaviors near $r = r_\odot$ can exhibit very different behaviors at large heliocentric distances. In fact, the present system (more complex than that discussed in I.4 because of the explicit presence of the energy equation) has recently been shown to possess different classes of solutions in which $T \to 0$ as $r \to \infty$ in significantly different manners.

Class 1: Solutions in which $T(r)$ approaches the form $T \propto r^{-2/7}$ as $r \to \infty$. The ratio ε of enthalpy flux and heat conduction flux $F_c = 4\pi r^2 f_c$ is

$$\varepsilon = \frac{4\pi n u r^2 \cdot 5kT}{4\pi r^2 \kappa_0 T^{\frac{5}{2}} \dfrac{dT}{dr}}. \tag{3.13}$$

For $T \propto r^{-2/7}$, the heat conduction flux takes on a constant, finite value, $F_{c\infty}$, while the enthalpy flux approaches zero, as $r \to \infty$. Thus $\varepsilon \to 0$. Examination of the total energy flux integral (3.12) reveals that at large r, the expansion speed u must approach a limiting value $u_{\infty 1} = [2(F - F_{c\infty})/I]^{1/2}$. Solutions of the one-fluid equations with this behavior of $T(r)$ as $r \to \infty$ have been derived by Parker [3.14] and by Noble and Scarf [3.15].

Class 2: Solutions in which $T(r)$ approaches the form $T \propto r^{-2/5}$ as $r \to \infty$. Both the enthalpiy flux and the heat conduction flux vary as $r^{-2/5}$ for $r \to \infty$, so that the ratio ε approaches a constant, finite value at large r. Examination of the energy flux integral reveals that, in contrast to Class 1 above, the expansion speed u must approach a limiting value $u_{\infty 2} = (2F/I)^{1/2}$ at large r. The solution of the one-fluid equations with this behavior of $T(r)$ as $r \to \infty$ has been derived by Whang and Chang [3.16], and will be described in greater detail below.

Class 3: Solutions in which $T(r)$ approaches the form $T \propto r^{-4/3}$ as $r \to \infty$. For $T \propto r^{-4/3}$, the heat conduction flux approaches zero more rapidly than the enthalpy flux. Thus $\varepsilon \to \infty$. Examination of the total energy flux integral (3.12) reveals that at large r, the expansion speed must approach a limiting value $u_{\infty 3} = (2F/I)^{1/2}$. This class of solutions appears not to have been anticipated until recently [3.17, 3.13]. Actual numerical solutions have been computed (and shown to pass through the critical point) by Durney [3.17], while the mathematical properties of the class have been explicated by Roberts and Soward [3.18]. Physically, $T \propto r^{-4/3}$ corresponds to an adiabatic expansion—i.e., $S(r)$ is negligible and $P \rho^{-\gamma}$ constant ($\gamma = 5/3$) at large heliocentric distances.

The precise relationship among these classes of possible solutions to the one-fluid equations has been somewhat obscure (e..g, see Parker [3.19]). Although a detailed discussion of this relationship is beyond the scope of our present discussion, it recently has been demonstrated [3.18] that for a given coronal temperature T_0, low coronal densities lead to the first type of solution ($T \propto r^{-2/7}$), while higher coronal densities lead to the third type of solution ($T \propto r^{-4/3}$). The second type of solution ($T \propto r^{-2/5}$) exists at a single intermediate density. For extremely high densities, the subsonic models of Chamberlain become the valid solutions to the one-fluid equations.

It might thus appear that three different types of "one-fluid" models are available for comparison with solar wind observations and with the other models to be described below. Yet, in practice, only the Whang and Chang model (with $T \propto r^{-2/5}$ at large r) can be so utilized. The solutions with $T \propto r^{-2/7}$ at large r have been approximated analytically under the assumption that the flux of kinetic energy, $F_\kappa = 4\pi n u r^2 \cdot \frac{1}{2} m u^2$, is everywhere much smaller than the heat conduction flux F_c [3.14]. Table 3.2 reveals this to be a poor approximation near 1 AU in the observed solar wind. The solutions with $T \propto r^{-4/3}$ at large r have not yet been published in sufficient detail to allow comparison with observations at a finite heliocentric position. Thus we must rely on the Whang and Chang model in the following discussions.

This practical limitation is more serious than may be at first apparent. The Whang and Chang solution, with $T \propto r^{-2/5}$ at large heliocentric distances, involves a restriction on the value of the ratio $\varepsilon(r)$ defined in equation (3.13); Whang and Chang demonstrated that a solar wind solution with this behavior of T at large r is a possible solution with this behavior of T at large r is possible only when $\varepsilon(r_c) = 1.26265$, where r_c is the critical radius. This restriction acts as an additional boundary condition and the boundary value problem, as stated above, would be overdetermined. A proper specification of boundary conditions for this class of solution then involves choice of only one of the "coronal conditions" n_0 and T_0; choice of one effectively determines a single permissible value of the other (as implied in the above description of the relationship among the different classes of solutions). The resulting solution is then a unique relationship between two independent variables, the dimensionless expansion speed $U = u/u_c$ and the dimensionless sound speed $A = \sqrt{T/T_c}$, and the single independent variable, the dimensionless heliocentric position $X = r/r_c$. Specification of the location of the critical radius r_c then determines the physical variables u, T, and n (through the mass flux integral) *at all heliocentric positions*. Choice of r_c to be in the range 5 to $10 r_\odot$ yields solutions that imply reasonable coronal temperatures, 1 to $2 \times 10^6 \, °K$, but rather low coronal densities, less than $10^8 \, cm^{-3}$, for r near $r_\odot$, the solar radius. The observed coronal densities of $\sim 10^8 \, cm^{-3}$ then indicate that the Whang and Chang model, drawn from the most restrictive class of one-fluid solutions, is only marginally applicable to the solar corona.

Table 3.3   *The Basic "One-Fluid" Model of the Coronal Expansion* $(r_c = 7.5 \, r_\odot)$

	$r = r_\odot$	$r = 1 \, AU$	$r \to \infty$
Density, cm^{-3}	7.4×10^7	8	
Expansion Speed, km sec^{-1}	1.2	260	315
Temperature, °K	1.6×10^6	1.6×10^5	$\alpha r^{-2/5}$

Table 3.3 summarizes the predictions of the Whang and Chang model, to be used (despite the above reservations) and referred to as the "basic one-fluid model" in the remainder of our discussion. The critical radius has been chosen to be at $7.5 \, r_\odot$, a choice judged by Whang and Chang to give the best agreement with coronal density observations. Selection of a smaller critical radius would imply a lower coronal density and higher coronal temperature, and would predict high density, flow speed, and temperature at $1 \, AU$. Selection of a larger critical radius would reverse all of these changes.

III.5 A "Two-Fluid" Model of the Steady, Spherically-Symmetric Coronal Expansion

Our entire previous discussion of coronal expansion models has been based on the assumption of equal proton and electron temperatures. This assumption is valid only if the two particle species interact strongly enough to maintain this equipartition of thermal energy. Deep in the corona, coulomb collisions between protons and electrons occur with sufficient frequency to insure equipartition. However, as a parcel of plasma moves away from the sun in the coronal expansion, the density decreases from $\sim 10^8 \, cm^{-3}$ near $r_\odot$ to only $\sim 10 \, cm^{-3}$ near $1 \, AU$. The coulomb collision frequency is proportional to the density, so that the drastic density change inherent in the expansion leads to very low collision frequencies in interplanetary space. Sturrock and Hartle [3.20] argued that this rate was so low that the protons and electrons were not constrained to have equal temperatures. Subsequent observations (see Table 3.1) have borne out their prediction. Incorporation of this non-equilibrium effect into the fluid equations describing the coronal expansion leaves the mass conservation law (3.1) unchanged, and requires only the use of the two-component pressure term

$$P = nk(T_e + T_p)$$

in the momentum equation (3.2). However, separate energy equations must be written for protons and electrons. In their formulation of this "two-fluid" model of the coronal expansion, Sturrock and Hartle used energy equations in an alternate form (obtained by multiplying the momentum equation 3.2 by u and subtracting from the energy equation 3.3) that eliminates the bulk motion and body forces:

$$nu\left(\frac{3}{2}k\frac{dT_p}{dr} - \frac{kT_p}{n}\frac{dn}{dr}\right) = \frac{1}{r^2}\frac{d}{dr}\left(r^2\kappa_p\frac{dT_p}{dr}\right) + \frac{3}{2}vnk(T_e - T_p) \qquad (3.14)$$

and

$$nu\left(\frac{3}{2}k\frac{dT_e}{dr} - \frac{kT_e}{n}\frac{dn}{dr}\right) = \frac{1}{r^2}\frac{d}{dr}\left(r^2\kappa_e\frac{dT_e}{dr}\right) - \frac{3}{2}vnk(T_e - T_p) \qquad (3.15)$$

for protons and electrons respectively. The energy source terms are due to heat conduction by each particle species and to energy exchange between the two species. The energy exchange term tends to eliminate any difference between T_p and T_e on the time scale $\tau = 1/v$. This exchange was assumed to occur by coulomb collisions, with the frequency

$$v = \frac{2m_e}{m_p}v_e \qquad (3.16)$$

where v_e is the electron collision frequency [3.21]. For our present purposes the energy exchange frequency can be written in the approximate form [3.22]

$$v = 9 \times 10^{-2} n T^{\frac{3}{2}}. \qquad (3.17)$$

Solar wind solutions of the system of "two-fluid" equations (3.1), (3.2), (3.14), and (3.15) were obtained numerically by Sturrock and Hartle. The boundary conditions involved specification of the density, proton temperature, and electron temperature at $r = r_0$, the requirement that the solution pass through the critical point, and the requirements that $T_e \to 0$ and $T_p \to 0$ as $r \to \infty$. The choice of $n_0 = 3 \times 10^7 \, \mathrm{cm}^{-3}$ and $T_{p0} = T_{e0} = 2 \times 10^6 \, ^\circ\mathrm{K}$ at $r_0 = r_\odot$ placed the critical radius at $r = 7.1 r_\odot$. Fig. 3.7 shows the predicted density as a function of heliocentric distance in the range $r_\odot < r \leq 10^4 r_\odot$, along with two sets of observed coronal densities [3.23, 3.24]. For $r_\odot \lesssim r < 2r_\odot$, the predicted densities are consistently and significantly lower than observed values (by a factor of ~3 at $r \approx r_\odot$). Hartle and Sturrock [3.22] attributed this difference to the presence of an energy source other than heat conduction in this part of the corona. In other words, the basic energy dissipation mechanism that produces the high coronal temperature is postulated to be important as far from the base of the corona as $r \approx 2r_{\odot}$. This interesting interpretation is not unique; the presence of strong coronal magnetic fields could have a similar effect on $n(r)$. For $2r_{\odot} \lesssim r \lesssim 20r_{\odot}$, the predicted and observed coronal densities agree reasonably well.

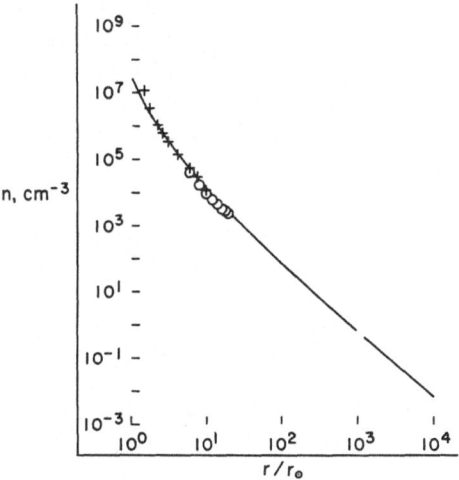

Fig. 3.7 The density $n(r)$ predicted by a two-fluid model of the coronal expansion [3.22], along with coronal densities observed by Blackwell [3.23] (as open circles) and by Michard [3.24] (as crosses)

We will utilize the two-fluid solution with coronal conditions specified above as our "basic two-fluid model" in subsequent discussions and comparisons with solar wind observations. Table 3.4 summarizes the predictions of this model. Fig. 3.8 shows the predicted temperatures $T_p(r)$ and $T_e(r)$ along with the single temperature $T(r)$ given by the one-fluid model (with its critical radius adjusted to also give $T(r_{\odot})$ $= 2 \times 10^{6}\,^{\circ}\text{K}$). The fundamental differences between the predictions of the

Table 3.4   The Basic "Two-Fluid" Model of the Coronal Expansion ($r_c = 7.1 r_{\odot}$)

	$r = r_{\odot}$	$r = 2r_{\odot}$	$r = 1\,\text{AU}$
Density, cm^{-3}	3×10^{7}	1.5×10^{6}	15
Expansion Speed, km sec^{-1}	5.8	29	250
Proton Temperature, $^{\circ}\text{K}$	2×10^{6}	1.2×10^{6}	4.4×10^{3}
Electron Temperature, $^{\circ}\text{K}$	2×10^{6}	1.5×10^{6}	3.4×10^{5}

two models are easily understood on the basis of the behavior of the heat conduction and energy exchange terms in equations (3.14) and (3.15). The coupling term in the two-fluid equations is (because of its dependence on n) a rapidly decreasing function of r. As the proton heat conductivity is small $(\kappa_p \approx \sqrt{m_e/m_p}\,\kappa_e,$ [3.21]) both sources in the proton energy equation (3.14) become small away from the sun, allowing the protons to cool nearly as rapidly as in an adiabatic expansion. In contrast, the heat conduction term in the electron energy equation (3.15) remains important throughout interplanetary space, keeping the electrons almost as hot as in an extended, static corona (I.3). The one-fluid model forces equipartition of the thermal energy stemming from electron heat conduction, and thus leads to a single temperature intermediate between the T_p and T_e of the two-fluid model.

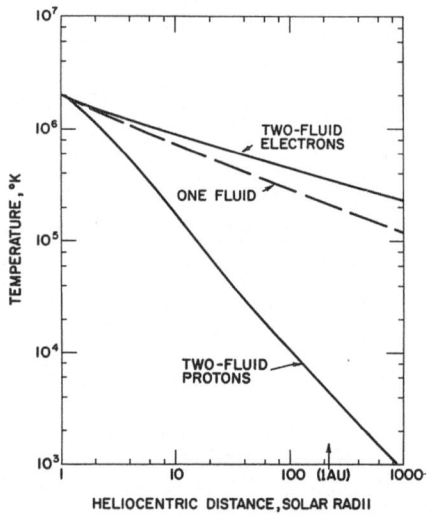

Fig. 3.8   The temperatures $T_e(r)$ and $T_p(r)$ predicted by the two-fluid model (solid lines) and the single temperature $T(r)$ predicted by the one-fluid model (dashed line) [3.25]. The same coronal temperature $T(r_\odot) = 2 \times 10^6\,^\circ\mathrm{K}$ has been assumed in both models

In addition to this easily-explained difference between the predicted temperature variations in the one-fluid and two-fluid models, another, more subtle difference also exists. The one-fluid model of Whang and Chang was chosen to give $T \propto r^{-2/5}$ as $r \to \infty$. Examination of Fig. 3.8 reveals [3.25, 3.26, 3.27] that for the two-fluid model of Sturrock and Hartle, $T_e \propto r^{-2/7}$ and $T_p \propto r^{-6/7}$ at large r. Thus the heat conduction flux approaches zero as $r \to \infty$ in the former model, but approaches a finite value, $F_{c\infty}$, as $r \to \infty$ in the latter model. We have seen in III.4 that three different classes of supersonic solutions, with different behaviors of $T(r)$ and $u(r)$ at large r, exist for the one-fluid equations. Is there a similar multiplicity of solutions to the two-fluid equations? Is it proper to compare one-fluid and two-fluid solutions with different boundary conditions as $r \to \infty$? Unfortunately, these questions have not been explored or answered. We will soon find that conclusions drawn in comparing the predictions of the two-fluid model with solar wind observations are clouded by such unanswered questions.

III.6 A Comparison of Solar Wind Observations with Predictions of the Basic One-Fluid and Two-Fluid Models

Table 3.5 summarizes the properties of the quiet solar wind, as provisionally identified with low flow speeds in III.2 and III.3, and the properties predicted by the basic one-fluid coronal expansion model of III.4 and the basic two-fluid coronal expansion model of III.5. It is abundantly clear that the predictions of these models and the observations made near 1 AU are in basic agreement. Our purpose here is a precise, rather than a general, comparison. In pursuing this end, we should be aware of two possible ambiguities. The first is the slender physical justification for our identification of the "quiet" or structureless state of the solar wind with low solar wind speeds, a point already belabored in III.2 and III.3. The second stems from the sensitivity of the solar wind properties predicted by these models to the assumed coronal density and temperature. The basic one-fluid and two-fluid models assume quite different coronal conditions, chosen, in fact, to optimize (in some sense) the agreement of each model with both coronal and solar wind observations.

Table 3.5 *A Comparison of Solar Wind Observations and Theoretical Models (at 1 AU)*

	Observed	One-Fluid Model	Two-Fluid Model
Density, cm^{-3}	9	8	15
Flow Speed, km sec^{-1}	320	260	250
Proton Temperature, °K	4×10^4	1.6×10^5	4.4×10^3
Electron Temperature, °K	1.5×10^5	1.6×10^5	3.4×10^5

The following conclusions can be drawn by comparing the observed and predicted solar wind properties as given in Table 3.5:

(1) The basic two-fluid model predicts a solar wind density about twice as high as observed, *despite* the choice of a coronal density (near

$r = r_\odot$) lower than indicated by coronal observations (III.5). The one-fluid model predicts a solar wind density in good agreement with observations with the choice of a higher coronal density than that used in the two-fluid model (but still somewhat lower than indicated by coronal observations).

(2) Both models predict solar wind speeds about 20% lower than the $\sim 320 \, km \, sec^{-1}$ value that we have indentified with the quiet solar wind. The solar wind speed is sometimes observed to be near $250 \, km \, sec^{-1}$, in agreement with the predictions of the models. However, such conditions are extremely rare (Fig. 3.6).

(3) The one-fluid model predicts a single temperature that agrees well with the observed electron temperature, but is about four times larger than the observed proton temperature. The ratio of electron and proton temperatures in the low-speed solar wind is ~ 4, showing that the assumption of equal temperatures implicit in the one-fluid model is invalid (to a moderate degree) near 1 AU.

(4) The two-fluid model predicts an electron temperature about three times higher than the observed value, and a proton temperature about an order of magnitude lower than the observed value. If this comparison were made using solar wind observations for which the flow speed is near $250 \, km \, sec^{-1}$, the electron temperature difference would be unchanged, while the predicted proton temperature is 3 to 5 times lower than the observed value [3.4, 3.25].

The accuracy of present day solar wind measurements of all quantities used in the comparisons made above appears to be sufficiently good that, aside from the problem of identification of the quiet solar wind, the above conclusions can be stated with assurance. The physical interpretation of these conclusions is much less certain. In fact, contrasting interpretations are central to much current discussion regarding the physical processes affecting the coronal expansion.

This discussion stems from two "problems" suggested by comparisons of observation and theory such as that given above. The first of these is the so-called "density problem," the tendency noted in comparison (1) for coronal expansion models to predict too high a ratio of the density at 1 AU to the coronal density. This leads to too high an interplanetary density if a density of $\sim 10^8 \, cm^{-3}$ near $r = r_\odot$, as indicated by coronal observations, is assumed. An interplanetary density in agreement with solar wind observations can be obtained only by using a coronal density somewhat lower than $10^8 \, cm^{-3}$. This problem is most severe for the basic two-fluid model, less clear for the basic one-fluid model. The second of these is the so-called "energy problem," the tendency noted in (2) for coronal expansion models to predict too low a solar wind speed (and thus, presumably, too low an energy in the flow) under the assumption of reasonable coronal densities and temperatures. This problem again appears to be more severe for the two-fluid model than for the one-fluid model; if the coronal density and temperature used in the basic one-fluid model (Table 3.3) are adopted in a two-fluid model, the latter gives an expansion speed of only 190 km sec^{-1} at 1 AU [3.22]. These two problems are not completely separable, as the density and flow speed are related directly by the particle flux integral (3.11) and thus indirectly by the general dynamics of the expansion.

Much of our forthcoming discussion will be organized in the context of the "energy problem." The implications regarding the "density problem" will be specifically mentioned where most appropriate. The reader should be forewarned that the problem of energetics in the coronal expansion is among the most controversial topics in current solar wind research, and one in which hard and firm conclusions have not been attained.

III.7 The Energetics of the Coronal Expansion

The problem of the energy source for the coronal expansion is an extension of the general problem of coronal heating. The existence of a 10^6 °K corona above the 6000 °K solar photosphere implies the existence of a physical mechanism that can transfer energy from a cold region to a hot region (unless the corona were heated from outside of the sun, a possibility now largely discredited). The mechanisms one would expect to be important in the outer layers of the sun, such as heat conduction and radiative transfer, cannot accomplish this feat. It is necessary to postulate some additional transport of energy out of the sun, through the photosphere, with dissipation of this energy on the corona (and the underlying chromosphere). It is now generally agreed that this additional energy is transported by waves generated in the turbulent convection zone lying just below the solar photosphere; those waves that can pass through the photosphere are damped in the lower density regions, the chromosphere and corona. There is much less agreement as to the nature of the waves and the mechanism for their damping (see the review by Kuperus [3.28]). Present-day coronal observations have provided little quantitative evidence regarding this heating mechanism.

It is possible, however, to estimate the energy losses from the layers of the solar atmosphere above the photosphere, and thus estimate the energy dissipation rate required to maintain the chromosphere and corona. Fig. 3.9 illustrates the resulting energy balance for (a) the entire outer solar atmosphere, and (b) the corona [3.28, 3.13]. The estimated energy losses are indicated along the right-hand side of the figure, and the energy inputs required to maintain an energy balance are indicated along the left-hand side. Losses from the entire system are dominated by radiation; even the corona loses most of its energy by conduction down into the chromosphere, from which the denser plasma can radiate more efficiently. The energy carried away in the coronal expansion is but a percent of the total loss from the entire region, and only about 10% of the total loss from the corona. Our discussion of the energy source for the coronal expansion should be regarded in this perspective.

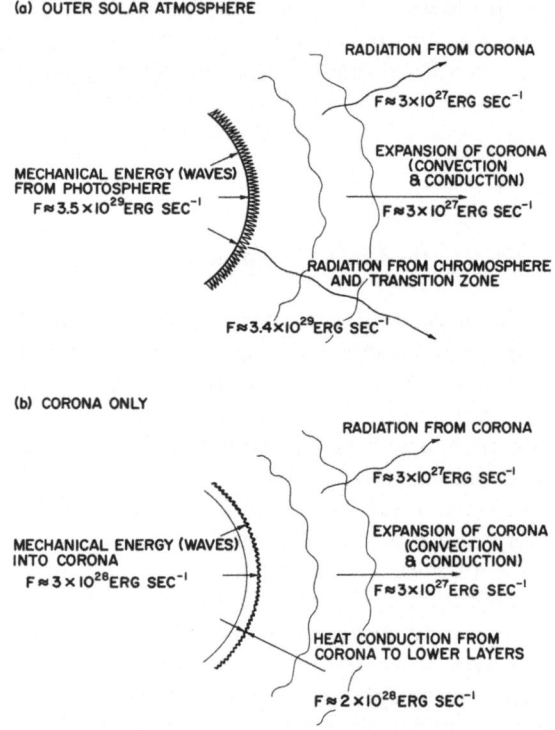

(a) OUTER SOLAR ATMOSPHERE

RADIATION FROM CORONA

$F \approx 3 \times 10^{27}$ ERG SEC^{-1}

MECHANICAL ENERGY (WAVES) FROM PHOTOSPHERE
$F \approx 3.5 \times 10^{29}$ ERG SEC^{-1}

EXPANSION OF CORONA (CONVECTION & CONDUCTION)
$F \approx 3 \times 10^{27}$ ERG SEC^{-1}

RADIATION FROM CHROMOSPHERE AND TRANSITION ZONE
$F \approx 3.4 \times 10^{29}$ ERG SEC^{-1}

(b) CORONA ONLY

RADIATION FROM CORONA

$F \approx 3 \times 10^{27}$ ERG SEC^{-1}

MECHANICAL ENERGY (WAVES) INTO CORONA
$F \approx 3 \times 10^{28}$ ERG SEC^{-1}

EXPANSION OF CORONA (CONVECTION & CONDUCTION)
$F \approx 3 \times 10^{27}$ ERG SEC^{-1}

HEAT CONDUCTION FROM CORONA TO LOWER LAYERS
$F \approx 2 \times 10^{28}$ ERG SEC^{-1}

Fig. 3.9   Estimated energy balances for the outer solar atmosphere and corona [3.13]

Coronal observations appear to show [3.28] that the maximum coronal temperature is attained no farther than ~ 1 solar radius above the sharp transition from the chromosphere to the corona; i.e., somewhere in the range $r_\odot \leq r \leq 2r_\odot$. Beyond the temperature maximum, the corona might possibly be maintained against the energy losses due to expansion and radiation purely by heat conduction from below, without further significant energy dissipation. Whether this is in actuality the case could be determined by detailed knowledge of the coronal density and temperature structure. In the absence of such knowledge there arises interest in attempts to test the validity of the "conduction corona," with all energy sources other than heat conduction neglected beyond $r \approx r_{\odot}$, by indirect inference based on solar wind observations made at 1 AU.

The first discussion of the problem of solar wind energetics was given by Parker [3.19] on the basis of approximate analytic solutions of the one-fluid equations (III.4) derived under the assumption that the heat conduction and gravitational terms dominated the energy flux (3.12). Then, in some range of heliocentric distance $r > r_0$,

$$F \approx -4\pi n u r^2 \frac{G M_{\odot} m}{r^2} - 4\pi r^2 \kappa_0 T^{\frac{5}{2}} \frac{dT}{dr}.$$

This equation is easily integrated (recalling that $4\pi n u r^2 = I$, a constant), to give

$$T = T_0 \left[\frac{r_0}{r} \left(\frac{1 + \dfrac{r^2}{r}}{1 + \dfrac{r_2}{r_0}} \right) \right]^{\frac{2}{7}}, \tag{3.18}$$

where

$$r_2 = r_0 \frac{I}{2F} \frac{G M_{\odot} m}{r_0}.$$

Equation (3.18) shows that $T \propto r^{-4/7}$ for $r \ll r_2$ and $T \propto r^{-2/7}$ for $r \gg r_2$. Parker then attempted to reconcile an approximate integration of the momentum equation, under the temperature dependence (3.18), with both the early solar wind observations (from Explorer 10 and Mariner 2) indicating a typical solar wind speed of $\sim 400\,\mathrm{km\,sec}^{-1}$ and a typical density of $\sim 4\,\mathrm{cm}^{-3}$ near 1 AU, and the coronal observations indicating a temperature of $\sim 10^6\,^{\circ}\mathrm{K}$ and a density of $\sim 10^8\,\mathrm{cm}^{-3}$ near $r = r_{\odot}$. Assumption of $r_0 = r_{\odot}$ and use of $n_{\odot} = 10^8\,\mathrm{cm}^{-3}$ and $T_{\odot} = 10^6\,^{\circ}\mathrm{K}$ led to $u = 330\,\mathrm{km\,sec}^{-1}$ at 1 AU, lower than the typical observed value. In fact, u could be raised as high as $400\,\mathrm{km\,sec}^{-1}$ only by increasing $T_{\odot}$ and decreasing $n_{\odot}$ to $\sim 10^7\,\mathrm{cm}^{-3}$. In addition to invoking an unacceptably low coronal density, these choices also led to densities of $\sim 2\,\mathrm{cm}^{-3}$ near 1 AU. The solar wind and coronal observations appeared to be reconcilable (in this approximation) only if the temperature fell off less rapidly than $r^{-2/7}$, a condition that would arise if there were an additional energy source beyond $r = r_{\odot}$. Parker thus drew an important conclusion [3.19]: "The comparison shows that the solar corona is heated by wave dissipation, in addition to thermal conduction, for some distance into space, and that the conduction corona is probably not a very good model for the solar corona." The solar wind emanating from above active regions, with observed densities of $\sim 10^9\,\mathrm{cm}^{-3}$, was even more difficult to reconcile with the higher solar wind speeds observed by Mariner 2 and thought to be associated with solar activity (Chapter V).

Parker's arguments for an energy source term in addition to heat conduction can be questioned, in the light of present knowledge, from several points of view. His approximate temperature solution (3.18) limits discussion to only one of the three possible classes of solutions to the one-fluid equations, as described in III.4. The validity of a conduction corona in the context of the other classes was not considered. In fact, the one-fluid model appropriate to a coronal temperature near $10^6\,°\mathrm{K}$ and a coronal density near $10^8\,\mathrm{cm}^{-3}$ appears to be in the class for which $T \propto r^{-4/3}$ at large r. Most important, the use of a model assuming a steady, spherically-symmetric flow in comparison with "typical" solar wind properties appears to be unjustified on the basis of the recent autocorrelation analysis of observed solar wind speeds (III.2). In fact, the value $u = 330\,\mathrm{km\,sec}^{-1}$ derived by Parker does not lead to an obvious contradiction when compared to the expansion speed used for the "quiet" solar wind in III.6. Finally, the use of a steady, spherically-symmetric model in comparison with elevated solar wind speeds would appear to be still less justified. Nevertheless, Parker's hypothesis of energy dissipation well out from the base of the corona is central to much of the subsequent discussion of solar wind energetics.

The development of the two-fluid model of the coronal expansion led to a refinement and extension of the energy dissipation hypothesis. This model again predicted too low an expansion speed (III. 6), even when a coronal density as low as $3 \times 10^7\,\mathrm{cm}^{-3}$ was assumed near $r = r_\odot$. The prediction of a proton temperature an order of magnitude lower than the observed value was also interpreted as evidence of a need for additional energy. These two pieces of evidence led Hartle and Sturrock [3.22] to a conclusion similar to that of Parker: "These considerations lead one inexorably [sic!] to the view that the solar wind is subject to a heating mechanism other than, and in addition to, that of thermal conduction. This conclusion may alternatively be expressed as the statement that *the non-thermal heating mechanism which is responsible for the high temperature of the corona is operative also in the solar wind*." This heating is in addition to that inferred from the difficulty in matching observed densities for $r \lesssim 2r_\odot$ (III.5).

We may question the arguments of Hartle and Sturrock on much the same grounds as we questioned those of Parker. Is the Hartle and Sturrock model with $T_e \propto r^{-2/7}$ as $r \to \infty$ the only possible class of solar wind solutions to the two-fluid equations? Are the solutions applicable at higher coronal densities of the form $T_e \propto r^{-4/3}$, as in the one-fluid case? Despite these unanswered questions, the concept of extended mechanical dissipation in the context of the two-fluid model has won considerable acceptance [3.29, 3.30].

This acceptance has not been universal. Observations of solar wind electron temperatures first became available just as Hartle and Sturrock formulated their arguments for "non-thermal heating." We have already seen (III.6) that the two-fluid model predicts an electron temperature about twice as high as the observed value (now quite well established). Although the difference of a factor of two in electron temperatures might at first seem less important than the order of magnitude difference in proton temperatures, more careful consideration reveals two arguments to the contrary. First, solar wind electrons are several times hotter than the protons, and the *energy* differences revealed by the comparisons of electron and proton temperatures are nearly equal and opposite. A mere redistribution of the thermal energy predicted by the two-fluid model could conceivably bring *both* the electron and proton temperatures into better agreement with observations [3.31]. Second, the transport of energy by heat conduction, a process included in the two-fluid model by use of the conventional $T^{5/2}$ conductivity law (3.7) but not considered in the arguments for non-thermal heating, is extremely sensitive to the electron temperature.

A correct basis for comparing the energetics of a coronal expansion model and actual solar wind observations is the energy flux integral analogous to equation (3.12) for the one-fluid case [3.31]. In terms of flux densities,

$$nu\left[\frac{1}{2}mu^2 + \frac{5}{2}k(T_p + T_e) - \frac{GM_\odot m}{r}\right] - \kappa_e\frac{dT_e}{dr} - \kappa_p\frac{dT_p}{dr} = \frac{F}{4\pi r^2}, \quad (3.19)$$

where F is the total (constant) energy flux through a sun-centered sphere of radius r. Table 3.6 compares the energy flux density terms of

Table 3.6 *Predicted and Observed Energy Flux Densities (in ergs cm^{-2} sec^{-1}) at 1 AU*

	Two-Fluid Model	Observed
Convection of Kinetic Energy	0.20	0.22
Convection of Enthalpy (Electrons)	0.045	0.011
Convection of Enthalpy (Protons)	0.0006	0.005
Convection of Gravitational Energy	−0.005	−0.004
Heat Conduction by Electrons	0.29	0.007
Heat Conduction by Protons	$\sim 10^{-10}$	$\sim 10^{-5}$
TOTAL	0.53	0.24

(3.19) as predicted by the basic two-fluid model at 1 AU and as actually observed. The observations show that over 90% of the energy carried past 1 AU by the solar wind is in the form of kinetic energy. Only 3% of the total energy is carried by heat conduction. The average total energy flux density is 0.24 ergs cm^{-2} sec^{-1}. The basic two-fluid model predicts that only 38% of the energy carried past 1 AU is in the form of kinetic energy, while 55% of the total energy is carried by heat conduction. The predicted total energy flux density is 0.53 ergs cm^{-2} sec^{-1}. Thus the two-fluid model actually predicts twice the observed energy flow in the solar wind, but with a distribution of energy that is far different from that observed at 1 AU. On the basis of this comparison, it is less clear that energy should be added to the basic two-fluid model to improve agreement with observations.

Reexamination of Fig. 3.8 reveals that the electron temperature $T_e(r)$ given by the two-fluid model is nearly proportional to $r^{-2/7}$ as close to the sun as $r \approx 10 r_\odot$. The large heat conduction flux is essentially constant beyond this heliocentric distance, with no conversion into kinetic energy. This property of the basic two-fluid model is probably related to the boundary condition on $T_e(r)$ at large r, and the questions raised at the end of III.5 and above regarding the existence of other classes of solutions again become relevant. The prediction of too large a heat conduction flux is, of course, intimately related to the prediction of too high an electron temperature. As the heat conduction flux density $f_c = -\kappa_0 T_e^{5/2} dT_e/dr$ is roughly proportional to $T_e^{7/2}$ at a given position, overestimation of T_e by a factor of 2.3 (as in Table 3.5) could be expected to lead to overestimation of f_c by a factor of ~ 20 (as in Table 3.6).

These considerations have led to attempts to improve the predictions of steady, spherically-symmetric models of the coronal expansion by means other than energy addition. Such attempts have followed two differents paths. One involves the recognition of the many idealizations (e.g., neglect of magnetic forces, viscosity, and the tensor nature of the pressure and the heat conduction flux in a magnetized plasma) implicit in the basic one-fluid and two-fluid models, and incorporates one or more of these effects into more realistic models. The second path involves the recognition that the transport coefficients and exchange rates (e.g., thermal conductivity and the electron-proton energy exchange frequency) used in the basic models were based on the assumption of frequent coulomb collisions—an assumption well known to be invalid in much of the expanding corona. This path then involves the incorporation of modified transport coefficients and exchange rates into new models.

Our forthcoming survey of the important extensions of Parker's original coronal expansion theory will be organized in the context of these different approaches to the problem of solar wind energetics.

In III.8, three formulations of new coronal expansion models, all of which assume an appreciable transport of energy through the lower corona in the form of hydromagnetic waves, will be discussed. In III.9—III.11, three different refinements, all of which involve addition of well-known forces to the momentum equation will be described. In III.12 and III.13 modifications of the energy transport or exchange processes invoked in the basic models will be considered. In III.14, the overall problem of energetics will be reconsidered in the light of these new models.

III.8 The Effects of Hydromagnetic Waves on the Coronal Expansion

The hypothesis that the coronal heating mechanism is operative well out into the solar corona (beyond ~ 2 solar radii) is entirely reasonable from a physical point of view. The energy required to produce the coronal expansion was shown in III.7 to be only $\sim 10\%$ of the energy required to maintain the entire corona. There is no compelling reason why the waves postulated to heat the corona could not carry this small amount of energy to $r \gtrsim 2r_\odot$ before being dissipated. For example, Barnes [3.33] has considered the propagation of hydromagnetic waves (i.e., waves with frequencies small compared to the proton gyrofrequency and wavelengths large compared to the proton or electron gyroradii) in the magnetized coronal and interplanetary plasma. Of the wave modes possible in such a medium, the fast, magnetosonic waves were found to dissipate most of their energy within a few solar radii of the base of the corona, but to also be capable of transporting some energy as far from the sun as 10 to $20 r_\odot$ before dissipation. *If* a sufficient efflux of such waves were present in the lower corona (it can be argued that fast magnetosonic waves cannot propagate into the corona from the photosphere [3.28, 3.34]), both an additional pressure (due to the energy density of the waves) and an additional energy source (due to dissipation) would be present.

The first quantitative incorporation of an extended heating source into a coronal model was carried out by Hartle and Barnes [3.27]. The energy equations (3.14) and (3.15) of the basic two-fluid model were modified to

$$nu\left(\frac{3}{2}k\frac{dT_p}{dr} - \frac{kT_p}{n}\frac{dn}{dr}\right) = \frac{1}{r^2}\frac{d}{dr}\left(r^2\kappa_p\frac{dT_p}{dr}\right) + \frac{3}{2}vnk(T_e - T_p) + \mathscr{P}_p, \quad (3.20)$$

and

$$nu\left(\frac{3}{2}k\frac{dT_e}{dr} - \frac{kT_e}{n}\frac{dn}{dr}\right) = \frac{1}{r^2}\frac{d}{dr}\left(r^2\kappa_e\frac{dT_e}{dr}\right) - \frac{3}{2}vnk(T_e - T_p) + \mathscr{P}_e. \quad (3.21)$$

where $\mathscr{P}_p$ and $\mathscr{P}_e$ are the additional energy source terms (in units of ergs cm^{-3} sec^{-1}) for the protons and electrons. It was assumed that $\mathscr{P}_e = 0$ and that

$$\mathscr{P}_p = D_0 \left(\frac{n}{n_0}\right) \exp\left[-\frac{\left(\frac{r}{r_{\odot}} - a\right)^2}{b^2}\right], \qquad (3.22)$$

where D_0, a, and b are constants that determine the strength, location, and spread of the energy source. The form of (3.22) has no physical significance, but was selected to simplify the subsequent analysis. The source term can be thought of as a result of the hypothetical damping of waves, but is not limited to this interpretation. The restriction of the added energy dissipation to the proton energy equation was apparently made in hopes of elevating the solar wind proton temperature above that predicted by the basic two-fluid model, but without elevating the already too high electron temperature.

The mass and momentum equations of the basic two-fluid model and the new energy equations (3.20) and (3.21) were integrated over the range $r \geq r_0 = 2r_{\odot}$ under the same boundary conditions as used in the basic two-fluid model. Conditions at the inner boundary were taken to be $n_0 = 1.5 \times 10^6$ cm^{-3}, $T_{e0} = 1.5 \times 10^6$ °K, and $T_{p0} = 1.2 \times 10^6$ °K (as in

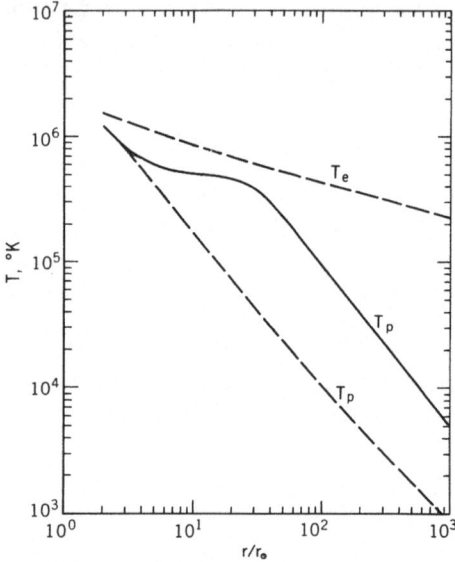

Fig. 3.10 The temperatures $T_e(r)$ and $T_p(r)$ predicted by the basic two-fluid model (dashed lines) and the $T_p(r)$ predicted by two-fluid model incorporating an extended energy source (solid line) [3.27]

Table 3.4). Fig. 3.10 shows the modification of the basic two-fluid solution (the dashed lines) for $T_e(r)$ and $T_p(r)$ produced by an energy source with $a=2$ and $b=22$, i.e., a source with maximum energy addition at the inner boundary and extending to $\sim 22 r_\odot$. The proton temperature is elevated most noticeably in the range $10 r_\odot \lesssim r \lesssim 20 r_\odot$, with a nearly adiabatic decrease beyond the latter position. Table 3.7 summarizes the predictions of this particular model. The proton temperature obtained at 1 AU is now in reasonable agreement with the observed value. The expansion speed has been elevated to $320 \, \text{km sec}^{-1}$ by the extended heating term and is in agreement with the solar wind speed taken to be indicative of quiet conditions. The density and electron temperature have been lowered slightly, remaining about twice the observed values. The problems with the heat conduction and total energy flux (III.7) are little changed.

Table 3.7 *A Two-Fluid Model with an Extended "Non-Thermal" Energy Source*

	$r=2r_\odot$	$r=1\,\text{AU}$
Density, cm^{-3}	1.5×10^6	13
Flow Speed, km sec^{-1}	2.5	320
Proton Temperature, °K	1.2×10^6	3.7×10^4
Electron Temperature, °K	1.5×10^6	3.4×10^5

The results of this particular model hold for more general choices of the parameters in the source function. Hartle and Barnes have thus confirmed the basic supposition of the "non-thermal heating" hypothesis—that an extended energy source can improve the agreement between the flow speeds and proton temperatures predicted by the two-fluid model and those actually observed in the solar wind. In fact, by suitable choices of the parameters D_0, a, and b, a sequence of models (with the same coronal conditions at $r=r_0$) resulted that give proton temperatures and flow speeds in agreement with observations over the range $250 \lesssim u \lesssim 400$ km sec^{-1}. We will comment on the implications of this result in Chapter 5.

Barnes, Hartle, and Bredekamp [3.35] have extended the above model by replacing the arbitrary form for $\mathscr{P}_p(r)$ given in (3.22) with the physical dissipation function expected for a given flux F_w of hydromagnetic waves propagating outward from $r_0 = 2r_\odot$. The energy source term $\mathscr{P}_e(r)$ in (3.21) was assumed to be negligible. The coronal density at $r=r_0$ was again taken to be $n_0 = 1.5 \times 10^6 \, \text{cm}^{-3}$, but new temperature values $T_{e0} = 1.3 \times 10^6 \, °\text{K}$ and $T_{p0} = 1.7 \times 10^6 \, °\text{K}$ were used. The model resulting from a flux $F_w = 2.1 \times 10^{26} \, \text{ergs sec}^{-1}$ of waves with a period of

Table 3.8   *A Two-Fluid Model with Wave Dissipation*

	$r=2r_\odot$	$r=1\,\text{AU}$
Density, cm^{-3}	1.5×10^6	15
Flow Speed, km sec^{-1}	2.9	330
Proton Temperature, °K	1.7×10^6	3.2×10^4
Electron Temperature, °K	1.3×10^6	2.2×10^5

5.2 min is summarized in Table 3.8. The flow speed and proton temperature predicted at 1 AU are again raised above the values from the basic two-fluid model to give reasonable agreement with observations. The electron temperature has been lowered significantly, and is now only 40% higher than the observed value. The density is little changed, still about twice that observed in the low speed solar wind.

This model removes some of the arbitrariness in the choice of $\mathscr{P}_p(r)$ made by Hartle and Barnes by relating the dissipation function specifically to a flux of hydromagnetic waves. It improves the predictions of the basic two-fluid model in much the same manner as the Hartle and Barnes model. The flux in waves of $2.1 \times 10^{26}\,\text{erg sec}^{-1}$ implies that $\sim 10\%$ of the energy in the solar wind is added by wave dissipation beyond 2 solar radii. The wave flux is only a few percent of that required to heat the corona. Variation of F_w, with other coronal conditions fixed, again leads to a sequence of models with a proton temperature that varies with flow speed in a manner similar to that actually observed. Discussion of this result (and its relevance to solar wind variations) will again be deferred to Chapter V.

The pressure associated with magnetohydrodynamic waves has been incorporated into quantitative models by Belcher [3.36] and by Alazraki and Couturier [3.37], for the case of linearly polarized Alfvén waves propagating outward from the sun along radial magnetic field lines. In the approximation that the expansion is polytropic (as in I.4), the total energy flux through a sun-centered sphere of radius r is (assuming spherical symmetry in the plasma flow)

$$F = 4\pi r^2 \left\{ nu \left[\frac{1}{2} mu^2 + \frac{\alpha}{\alpha - 1} \frac{P}{\rho} - \frac{GM_\odot m}{r} \right] + \frac{1}{2} \left(\frac{1}{2} \rho v^2 \right) u + \frac{1}{2} \frac{b^2}{4\pi} (u + c_A) \right\},$$

$$(3.23)$$

where α is the polytropic index, v^2 and b^2 are the mean-square amplitudes of the velocity and magnetic perturbations in the Alfvén waves, and c_A is the Alfvén speed. The last two terms of (3.23) give the energy flux associated with the waves. If these two terms are decreasing functions of heliocentric distance, wave energy is converted to kinetic and thermal energy in the expanding plasma. This conversion simply reflects the work done by the gradient in the wave pressure. Its presence is independent of any dissipation or damping of the waves, which would require an additional energy flux term in (3.23).

The amplitudes of an Alfvén wave propagating outward from the sun would change because of the variations in "background" plasma properties. These changes can be quantitatively analyzed using a "WKB" approximation [3.19]. Both v^2 and b^2 are found to decrease significantly between the corona and 1 AU (for example, $u \gg c_A$, as in interplanetary space, implies $b \propto \rho^{3/4}$, or $r^{-3/2}$). As a consequence, Alfvén waves propagating outward from the corona will do work (through the pressure gradient) on the solar wind plasma. Because the Alfvén speed is high in the corona, even small amplitude waves imply an appreciable pressure or energy flux; if the coronal density and magnetic field intensity are assumed to be $2 \times 10^7 \, \mathrm{cm}^{-3}$ and 1 gauss, the Alfvén speed at $r = 10^{11} \, \mathrm{cm}$ is $\sim 500 \, \mathrm{km \, sec}^{-1}$, and a wave amplitude of 0.15 gauss would imply an energy flux of $\sim 5 \times 10^{27} \, \mathrm{ergs \, sec}^{-1}$, essentially the total energy flux in the coronal expansion. Such an Alfvén wave pressure would then be expected to have a major effect on the expansion. For example, Belcher has added this flux to a polytropic expansion of a corona at $T_\odot = 1.7 \times 10^6 \, {}^\circ \mathrm{K}$. The solar wind speed (at 1 AU) was elevated from $\sim 160 \, \mathrm{km \, sec}^{-1}$ to $\sim 360 \, \mathrm{km \, sec}^{-1}$ by the wave pressure.

The relationship between these pressure effects and the energy dissipation discussed earlier is not entirely clear. If Alfvén waves are produced in the sun, they (unlike the fast magnetosonic waves) could propagate with little attenuation through the corona into interplanetary space (where they are observed, as described in II.3). Despite the decrease in wave amplitude with heliocentric distance, the relative amplitude, b/B, is found to increase. When $b/B \to 1$, nonlinear effects (so-called mode coupling) could convert some of the energy in Alfvén waves into the fast, magnetosonic waves considered by Barnes. If this process were to occur near the base of the corona, the flux of magnetosonic waves required in the energy dissipation theories could be produced. If this process were to occur well out in the corona or interplanetary space, the subsequent dissipation would have little further effect on solar wind dynamics, and the acceleration of the plasma by Alfvén wave pressure would be the dominant effect.

The relevance of these concepts to the corona remains debatable. Attempts to detect coronal wave motions through the Doppler broadening and shifting of emission lines [3.38, 3.39] have revealed no plasma motions with amplitudes as large as $\sim 2 \, \mathrm{km \, sec}^{-1}$. Unless the Alfvén waves have frequencies so high as to escape detection by this technique (oscillations with the five-minute period of photospheric and chromosphere disturbances would probably be detected), the limit on the wave amplitude would rule out energy fluxes greater than $\sim 5 \times 10^{24}$ ergs sec^{-1} in the low corona. Fluxes consistent with this limit could have no great effect on the coronal expansion. In contrast, Alfvén waves with a flux of $\sim 10^{27}$ ergs sec^{-1}, necessary to account for a large fraction of the observed solar wind energy flux, would imply plasma velocity amplitudes of ~ 50 km sec^{-1}. Such amplitudes (again excluding high frequencies) appear to be inconsistent with the coronal emission line studies.

III.9 The Effects of Magnetic Forces on the Coronal Expansion

In the presence of a magnetic field, a plasma is subject to the body force

$$\mathscr{F} = \frac{1}{c} \boldsymbol{j} \times \boldsymbol{B},$$

where $\boldsymbol{j}$ is the electrical current density and c is the speed of light. For hydromagnetic phenomena the displacement current can be neglected, so that $\boldsymbol{j} = (c/4\pi) \nabla \times \boldsymbol{B}$, and the magnetic force is

$$\mathscr{F} = \frac{1}{4\pi} (\nabla \times \boldsymbol{B}) \times \boldsymbol{B}.$$

We have already seen (I.5) that the coronal and interplanetary plasma is pervaded by a frozen-in magnetic field of solar origin, formed into a spiral configuration by solar rotation. As $\boldsymbol{B}$ has both radial and nonradial components, the magnetic force will not be radial even in a spherically symmetric flow with a uniform B (or a monopole field) at the base of the corona. Thus both radial and "azimuthal" (in the direction of solar longitude) momentum conservation equations must be solved. The small nonradial velocity component produced by the magnetic force will be discussed in III.15; we will be concerned here only with the changes in the radial expansion produced by the magnetic force.

The first coronal expansion model to include a magnetic force was a one-fluid, polytropic model formulated by Weber and Davis [3.40]. The topology of the solutions of the resulting radial momentum equation is still more complex than that considered in I.4; the solution analogous to Parker's solar wind solutions must pass continuously through *three* critical points associated with the sound speed (as in the nonmagnetic case), the Alfvén speed $c_A^2 = B^2/4\pi\rho$, and the "radial Alfvén speed" $c_{rA}^2 = B_r^2/4\pi\rho$, where B_r is the radial magnetic field component. The solutions obtained by Weber and Davis for the expansion speed $u(r)$ differed very little from those found by Parker for the nonmagnetic case; the magnetic force had little effect on the coronal expansion.

The magnetic force has also been included in one-fluid models by Urch [3.41], Brandt, Wolff, and Cassinelli [3.42], Weber and Davis [3.48], and Whang [3.43], and in a two-fluid model by Wolff, Brandt, and Southwick [3.44] (all assuming a monopole field at the base of the corona). Although the magnetic effect upon the expansion has never been found to be large, it has not been negligible in those models that include specific integration of the energy equation (rather than assuming a polytropic law). In all such cases, the solar wind speed obtained at 1 AU has been higher than in the basic one- and two-fluid models. Whang [3.43] has specifically pointed out that the radial component of the Poynting vector, as would appear in a hydromagnetic expression for the total energy flux, is appreciable near $r = 2r_\odot$ and decreases with heliocentric distance. This directly implies a significant conversion of magnetic energy into kinetic and thermal energies in the coronal expansion. In his quantitative, one-fluid model, Whang found that the magnetic force produced a 17% increase in the solar wind speed. Thus inclusion of the magnetic force could remove much of the difference between the expansion speeds predicted by the basic models and those observed in the "quiet" solar wind.

III.10 The Effects of the Magnetic Modification of Heat Conduction on the Coronal Expansion

The presence of a magnetic field can also significantly modify transport processes in a plasma. In particular, the thermal conductivity transverse to magnetic field lines is reduced by the approximate factor $1 + (\omega_g \tau_c)^2$, where ω_g is the gyrofrequency and τ_c is the mean time between collisions. This factor represents a reduction of the mean free path transverse to the magnetic field, with the gyroradius as the limiting path length in a strong field. In either the corona or interplanetary space, $\omega_g \tau_c \approx 10^5$, and the resulting thermal conductivity normal to the field lines is reduced by the factor of $\sim 10^{10}$. This inhibition of heat conduction has very little effect on a structureless coronal expansion near the sun, where the magnetic field lines are nearly radial, but becomes important in reducing the radial heat conduction flux at large heliocentric distances, where the spiral interplanetary magnetic field lines (Fig. 1.4) become tightly wound. Fig. 3.11 shows the conventional field line geometry near the solar equatorial plane. At heliocentric position r, the heat conduction flux density parallel to a magnetic field line is still given by

$$f_B = -\kappa_e \frac{d T_e}{d s},$$

where s is the distance along the field line. As the conduction flux density transverse to the field is negligible, the radial flux density is

$$f_r \approx f_B \cos \Phi ,$$

where Φ is the angle between the magnetic field and the radial from the sun. As $dr = ds \cos \Phi$,

$$f_r = - \kappa_e \cos^2 \Phi \frac{dT_e}{dr} .$$

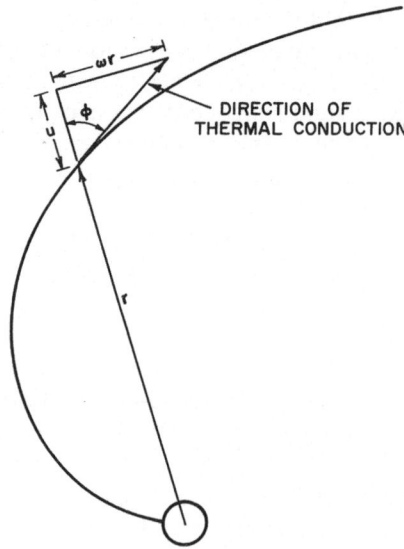

Fig. 3.11   The geometry of heat conduction in the presence of an interplanetary magnetic field

Thus the heat conduction flux density is reduced by the factor $\cos^2 \Phi$. Inspection of Fig. 3.11 reveals that

$$\cos^2 \Phi = \frac{1}{1 + \left(\dfrac{\omega r}{u} \right)^2} , \tag{3.24}$$

where ω is the angular velocity of the sun. The energy source term due to heat condution is then (near the solar equatorial plane)

$$S(r) = \frac{1}{r^2} \frac{d}{dr} \left(\kappa_0 \cos^2 \Phi \, T_e^{\frac{5}{2}} \frac{dT_e}{dr} \right) . \tag{3.25}$$

This effect has been incorporated into both one-fluid [3.41, 3.42, 3.43, 3.45] and two-fluid [3.44] coronal expansion models. The most important effect of the magnetic inhibition of radial heat conduction relates to the behavior of the solutions at large heliocentric distances, where $u(r)$ is nearly constant and

$$\cos^2 \Phi \approx \left(\frac{u}{\omega r}\right)^2 .$$

The heat conduction flux through a narrow band of solar latitudes subtending solid angle Ω near the solar equator is

$$F_c = \Omega r^2 f_r \rightarrow \Omega \left(\frac{u}{\omega}\right)^2 \kappa_0 T_e^{\frac{5}{2}} \frac{dT_e}{dr}$$

as $r \rightarrow \infty$. For *any* $T_e(r)$ that approaches zero as $r \rightarrow \infty$, the heat conduction flux approaches zero. This implies that the source term (3.25) becomes negligible at large r, leading to an adiabatic expansion with $T_e \propto r^{-4/3}$.

In a one-fluid formulation, this modified heat conduction law removes the multiplicity of possible forms for $T(r)$ at large r that was discussed in III.4. In the notation of that earlier discussion, ε must become infinite as $r \rightarrow \infty$. Further, the solutions including the magnetic inhibition factor are not "singular" in the manner of the basic (Whang and Chang) one-fluid model, but exist at a given coronal temperature $T_\odot$ for a wide range of coronal densities $n_\odot$. Fig. 3.12 compares the $T(r)$

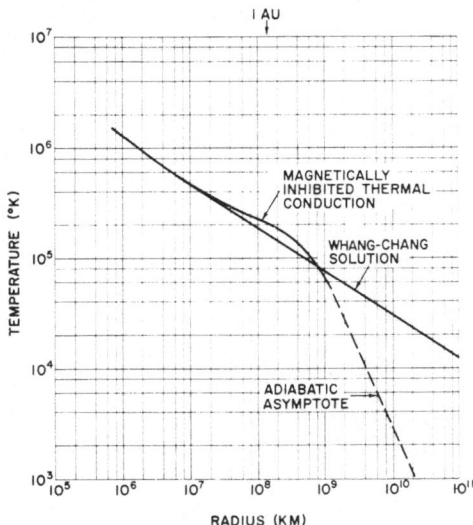

Fig. 3.12 The temperature $T(r)$ predicted by the basic one-fluid model and a one-fluid model with magnetic inhibition of heat conduction across magnetic field lines

from the basic one-fluid model with that from an "inhibited conduction" model (hereafter referred to as the "IC" model) [3.45] with the same choice of $n_\odot$ and $T_\odot$. The temperature profiles are nearly identical until $r \to 1\,\text{AU}$ (where the factor $\omega r/u$ in 3.24 becomes comparable to one); the approach to an adiabatic temperature law for the IC model occurs beyond this radius. Table 3.9 compares the solar wind properties predicted at 1 AU by the basic two-fluid model and by the one-fluid, IC model with the same coronal temperature and density. The latter predicts a significantly higher expansion speed, undoubtedly because of the more complete conversion of the total energy flux into kinetic energy flux (III.7). Table 3.10 summarizes the IC model chosen to give best agreement with solar wind observations at 1 AU . The only remaining disagreement is in the temperature. Although the coronal density dictated by this choice is lower than the accepted value (as in the basic two-fluid model) it would appear that this model can lead to higher expansion speeds and lower interplanetary densities than either the basic one-fluid model (with its singular nature) or the basic two-fluid model (with its finite heat conduction flux at large r).

Table 3.9 *A Comparison of the Solar Wind Properties at* 1 AU *Predicted by the Basic Two-Fluid Model and the One-Fluid Model with Magnetic Inhibition of the Heat Conduction*

	Two-Fluid	Magnetically-Inhibited One-Fluid
Density, cm^{-3}	15	14
Flow Speed, km sec^{-1}	250	310
Proton Temperature, °K	4.4×10^3	1.5×10^5
Electron Temperature, °K	3.4×10^5	1.5×10^5

Table 3.10 *A "Selected" One-Fluid Model with Magnetically-Inhibited Heat Conduction*

	$r = r_\odot$	$r = 1\,\text{AU}$
Density, cm^{-3}	3×10^7	8.4
Flow Speed, km sec^{-1}	4.3	324
Temperature, °K	1.8×10^6	2.8×10^5

Some additional understanding of the energetics of the coronal expansion can be gained by further examination of the one-fluid model with magnetically-inhibited heat conduction. Fig. 3.13 shows the variation with heliocentric distance of the various energy flux terms for the IC model (as in the energy conservation law 3.12). Near $r=r_\odot$ the heat conduction flux, $-4\pi r^2 \kappa_0 T^{5/2}(dT/dr)$, and the gravitational energy flux, $-4\pi n u r^2 (GM_\odot m/r)$, are nearly equal in magnitude but of opposite sign. Thus the next most important term, the enthalpy flux $4\pi n u r^2 \cdot 5kT$ accounts for nearly the entire total energy flux. This near cancellation of the two dominant fluxes, one of which is proportional to a high power of the coronal temperature, leads to a very sensitive dependence of the predicted solar wind properties upon coronal conditions. For example, at a coronal density of $1.2 \times 10^8\,\mathrm{cm}^{-3}$, a 25% change in the coronal temperature changes the expansion speed predicted at 1 AU by 20%, and the density predicted at 1 AU by a factor of six!

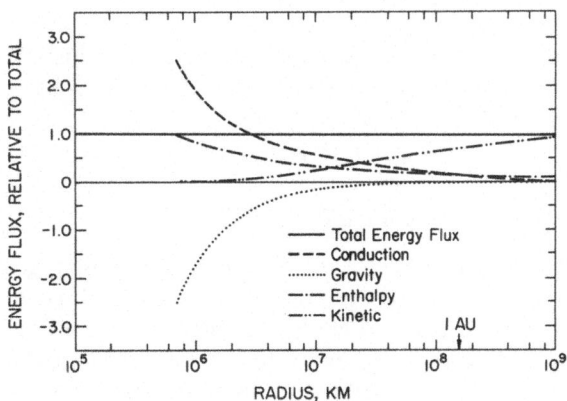

Fig. 3.13   Energy flux terms as functions of heliocentric distance for the one-fluid model with magnetic inhibition of heat conduction

III.11 The Effects of Viscosity on the Coronal Expansion

The radial velocity gradient present in a spherically-symmetric coronal expansion results in a viscous force that, in the absence of a magnetic field, has the radial component [3.46]

$$\mathscr{F}_r = \frac{1}{r^2}\frac{d}{dr}(r^2 Q_t).$$ (3.26)

The energy source terms are

$$S_p(r) = \frac{1}{r^2}\frac{d}{dr}(r^2 Q_p u)$$ (3.27)

and

$$S_e(r) = \frac{1}{r^2} \frac{d}{dr} (r^2 Q_e u) \qquad (3.28)$$

for a two-fluid model, and

$$S(r) = \frac{1}{r^2} \frac{d}{dr} (r^2 Q_t u) \qquad (3.29)$$

for a one-fluid model. The subscripts p and e refer to protons and electrons, Q denotes a viscous stress tensor, and $Q_t = Q_e + Q_p$. In the conventional hydrodynamic approximation, the viscous stresses are related to the velocity gradient by

$$Q_{p,e,t} = \frac{4}{3} \mu_{p,e,t} r \frac{d}{dr} \left(\frac{u}{r} \right),$$

where μ_p, μ_e, and μ_t are the viscosities. If coulomb collisions dominate interactions among the particles, the viscosities are given by the classical expressions [3.44]

$$\mu_p = 9.21 \times 10^{-17} T_p^{\frac{5}{2}},$$

and

$$\mu_e = 1.156 \times 10^{-18} T_e^{\frac{5}{2}}.$$

As $\mu_p \gg \mu_e$ at a given temperature, the proton viscosity generally has the more important influence on plasma dynamics, and $Q_t \approx Q_p$.

The viscous force (3.26) and energy source (3.29) were first applied to one-fluid expansion models by Scarf and Noble [3.47] and by Whang, Liu, and Chang [3.46]. The latter demonstrated that the presence of the viscous term removed the singularity (or "critical point" of I.4) in the momentum equation. Table 3.11 summarizes the resulting solar wind solution judged to give the best fit to coronal conditions [3.46]. The model predicts an expansion speed at 1 AU only half as large as the observed value, and in this respect can be regarded as a rather poor model of the coronal expansion. (Curiously, the density and temperature predictions are in good agreement with observations.) The viscosity used in this model has removed so much energy from the flow as to seriously decrease the coronal expansion speed.

Table 3.11   *A Nonmagnetic, Viscous Model of the Coronal Expansion*

	$r = r_\odot$	$r = 1\,\text{AU}$
Density, cm^{-3}	2×10^8	8.5
Flow Speed, km sec^{-1}	0.3	165
Temperature, °K	1.6×10^6	9×10^4

In actuality, the interplanetary magnetic field modifies the viscosity in a manner similar to that already described for the thermal conductivity. The viscous stress tensor becomes considerably more complicated [3.48], but to a reasonable approximation the relevant components become [3.44]

$$Q_{p,e} = \frac{4}{3} \mu_{p,e} r \frac{d}{dr}\left(\frac{u}{r}\right) \cos^2 \Phi, \qquad (3.30)$$

where Φ is again the angle between the magnetic field and the radial from the sun. The more complex and exact expressions have been incorporated into a one-fluid model by Weber and Davis [3.48], and the approximations (3.30) have been incorporated into a two-fluid model by Wolff, Brandt, and Southwick [3.44]. In both cases, the results are quite different from those of Whang, Liu, and Chang. The expansion is slowed to only a slight degree by the magnetically-modified viscosity; the most important dynamic effect is to enhance the nonradial component of the expansion velocity (see III.15). Wolff, Brandt, and Cassinelli have emphasized another important thermodynamic effect of viscosity in the two-fluid model. Fig. 3.14 shows the importance of the

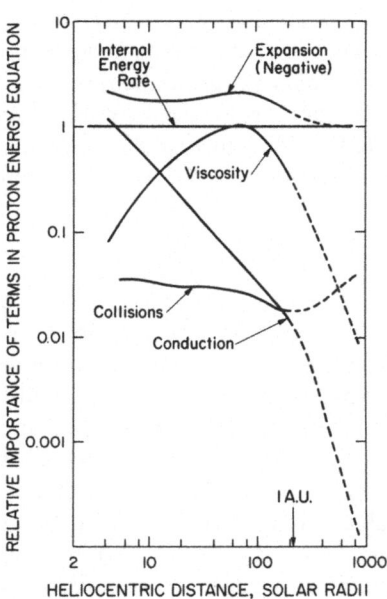

Fig. 3.14 The radial variations of the terms in the proton energy equation used in the two-fluid model of Wolff, Brandt, and Southwick [3.44]. All terms have been normalized to the local value of the internal energy rate of change $\frac{3}{2} nur^2 k(dT/dr)$

terms in the proton energy equation (relative to the rate of change of the internal energy density, $\frac{3}{2}nuk(dT_p/dr)$) as a function of heliocentric distance. In the range $10r_\odot \lesssim r \lesssim 600r_\odot = 3\,\mathrm{AU}$, the viscous heating term is the most important of all proton energy sources. Its effect alone was found to be sufficient to raise the proton temperature from the low level predicted by the basic two-fluid model to a value in agreement with solar wind observations.

III.12 The Effects of Noncollisional Energy Exchange Mechanisms on the Coronal Expansion

The exchange of energy between electrons and protons in the basic two-fluid model has been assumed to occur entirely by coulomb collisions. The rarity of such collisions in the tenuous plasma of the outer corona and interplanetary space leads to the large difference in the solar wind electron and proton temperatures predicted by the two-fluid model. It is possible in so complex a physical system as a plasma for other types of interactions to replace coulomb collisions as the primary means of energy exchange. The tendency of the basic two-fluid model to predict an electron temperature twice as high as observed, but a proton temperature much lower than observed, could be remedied by the presence of such a noncollisional exchange mechanism (strong enough to nearly, but not quite, partition the energy between the solar wind electrons and protons) [3.31, 3.10]. Such a mechanism can thus be hypothesized on empirical grounds similar to those for the hypotheses of "nonthermal heating" (III.7).

A physical justification for the existence of a noncollisional energy exchange mechanism can also be given. The coronal expansion drives the interplanetary plasma far away from the equilibrium state that would be maintained if collisions were frequent. The nonequilibrium nature of the plasma near 1 AU is attested by observations of differing electron and proton temperatures, of anisotropies in thermal motions and of the "skewing" of both electron and proton distribution functions associated with heat conduction. In the presence of such nonequilibrium features, a plasma is often unstable, capable of generating (internally) waves that can limit the deviations from equilibrium. Many such instabilities have been suggested for the interplanetary plasma [3.49], and some of these involve waves that can transfer energy from the hotter electrons to the cooler protons. The most promising of these appears to be a class of instabilities generated by skewing of the electron distribution function, as will be specifically discussed in III.13.

Noncollisional energy exchange rates have been incorporated into two-fluid solar wind models by several authors. Nishida [3.50] estimated that an energy exchange rate ~ 10 times faster than that for coulomb collisions was necessary to maintain the observed difference in proton and electron temperatures at 1 AU. Cuperman and Harten [3.51] integrated the two-fluid equations of motion using an energy coupling term $\frac{3}{2}\nu nk(T_e - T_p)$, as in (3.14) and (3.15), but where ν is some constant multiple ζ of the collisional frequency given in (3.16) and (3.17). A sequence of models was obtained, all with the same coronal conditions at a critical point near $r = 5.4 r_\odot$, but with different choices of ζ. As ζ was increased from the limiting value of 1 (pure collisional coupling), all solar wind properties predicted at 1 AU changed in the directions giving better agreement with observations. All of these changes were small except for the increase in the proton temperature. Table 3.12 summarizes the solar wind properties predicted at 1 AU by models with $\zeta = 1$ and $\zeta = 30$. The latter enhancement of the energy exchange rate brings the proton temperature into agreement with the observations. If ζ is increased still further, the model approaches a one-fluid model, as would be expected. Toichi [3.52] has integrated two-fluid energy equations, including a proton thermal anisotropy, assuming the radial expansion speed $u(r)$ derived in an earlier one-fluid model. An electron-proton energy exchange rate 10 to 100 times that given by coulomb collisions is again required to give the observed proton and electron temperatures at 1 AU.

Table 3.12 *Solar Wind Properties Predicted at 1 AU by Two-Fluid Models with Energy Exchange Rates Enhanced by the Factor ζ*

	$\zeta = 1$	$\zeta = 30$
Density, cm^{-3}	6.0	5.8
Flow Speed, km sec^{-1}	271	280
Proton Temperature, °K	1.6×10^4	4×10^4
Electron Temperature, °K	2.6×10^5	2.6×10^5

III.13 The Effects of a Reduced Thermal Conductivity on the Coronal Expansion

In all of the preceding discussion, the heat conduction flux density has been assumed to be proportional to the temperature gradient

$$f_c = -\kappa_e \frac{dT_e}{dr}, \tag{3.31}$$

with the proportionality constant κ_e that given by the classical theory [3.21] of transport phenomena in a collision dominated proton-electron gas;

$$\kappa_e = \kappa_0 \, T^{\frac{5}{2}}. \tag{3.32}$$

While the use of these concepts may be valid in the relatively dense solar corona, there is little justification for their application in the tenuous interplanetary plasma.

The first argument for questioning the application of (3.31) and (3.32) to the solar wind stems from the recognition that they are based on an analysis of *small* perturbations of an equilibrium (Maxwellian) distribution function in the presence of a temperature gradient. The dimensionless parameter

$$B_T = \frac{2 T_e}{\pi n q^4 \ln \Lambda} \frac{d T_e}{d r}$$

(where $\ln \Lambda$ is the so-called coulomb logarithm) plays a central role in this transport theory [3.53]. Both the proportionality of f_c and dT/dr, and the classical expression for the proportionality constant (thermal conductivity) hold only if $|B_T| \ll 1$. The physical interpretation of this restriction is simple. B_T is proportional to the ratio of the mean free path between collisions to the scale length of the temperature gradient. The requirement that $|B_T|$ be small assures that there are many collisions, tending to keep the distribution function nearly Maxwellian, in the distance that characterizes the temperature gradient. Fig. 3.15 shows $|B_T|$ as a function of heliocentric distances for the basic one-fluid and

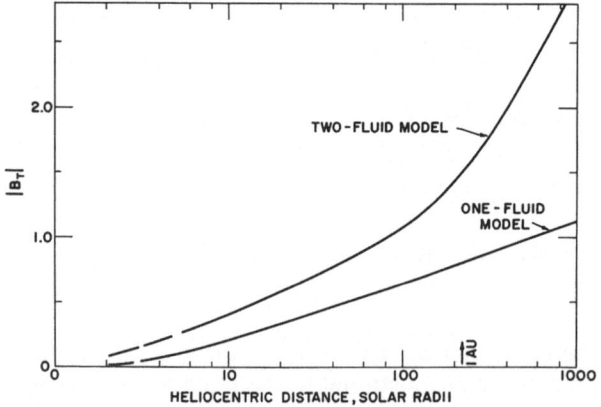

Fig. 3.15   The radial variations of the parameter B_T (see text) in the basic one-fluid and two-fluid models [3.13]

two-fluid models of the coronal expansion. The condition $|B_T| \ll 1$ is satisfied only for $r \lesssim 50 r_\odot$ in the one-fluid model, and $r \lesssim 15 r_\odot$ in the two-fluid model. The use of (3.31) and (3.32) is clearly not justified beyond these positions, i.e., in most of interplanetary space.

A second, related argument for questioning the applicability of (3.31) and (3.32) to the solar wind stems from the concept of a "saturation" level for the heat conduction flux density. It is evident [3.14] that internal energy cannot be transported by thermal motions at a rate greater than (or probably even as large as) the value obtained by assuming transport of the entire electron thermal energy $\frac{3}{2} n k T_e$ at the mean thermal speed of the electrons, $v_{th} = \sqrt{3 k T_e / m_e}$. Thus the heat conduction flux density should be limited by the saturation value

$$f_{cs} = \frac{1}{2} n m_e \left(\frac{3 k T_e}{m_e} \right)^{\frac{3}{2}} .$$

Note that f_{cs} is proportional to the density, while (3.31) and (3.32) imply a density-independent heat conduction. It is easily shown that use of the conventional plasma conductivity at large heliocentric distances leads to $f_c \rightarrow f_{cs}$ at large r unless the electron temperature declines faster than $r^{-\frac{1}{2}}$. As the temperature varies as $r^{-2/5}$ and $r^{-2/7}$ respectively for the basic one-fluid and two-fluid models, both employ physically unrealistic heat conduction fluxes at large r. In fact, of the three possible boundary conditions for one-fluid solutions discussed in III.4, only that with $T \propto r^{-4/3}$, the adiabatic expansion, is consistent with the restriction that f_c remain less than the saturation limit.

Both of these arguments show that the conventional concept of a heat conduction flux proportional to the temperature gradient may break down in the interplanetary plasma. As the condition $|B_T| \ll 1$ is violated well within 1 AU in the basic coronal expansion models, the consequences of this departure from near-equilibrium conditions could be of great significance in relating the predictions of models to observations near 1 AU. If particle interactions within the plasma continued to occur only by collisions, the transport of thermal energy in the solar wind could only be treated by a theory special to the coronal expansion, approaching exospheric theory at large heliocentric distances. However, Forslund [3.54] has shown that several plasma instabilities can arise when $|B_T|$ is as large as a few tenths. Fig. 3.16 illustrates the shape of the electron and proton distribution functions along the direction of the temperature gradient. The skewness of the electron distribution, produced by the temperature gradient and leading to a net transport of thermal energy or heat conduction, and the small thermal spread of the proton distribution combine to permit the growth of waves with phase velocities falling in the region of positive slope between the origin and the peak of the electron distribution. Acoustic, electrostatic, and magnetosonic wave modes can all (under differing plasma conditions) lead to instability. The nonlinear interaction of the resulting, internally generated waves with the plasma would be expected to limit the transport of electron thermal energy and to transfer energy from electrons to protons. The relevance to the present discussion (and that of III.12) should be clear.

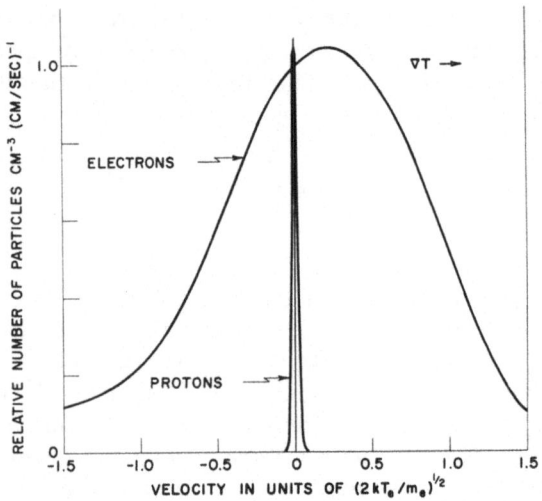

Fig. 3.16   Solar wind electron and proton velocity distribution functions, showing the skewness of the former produced by a temperature gradient

The development of a transport theory valid in the presence of the instabilities associated with finite B_T would be a formidable task, and has not been accomplished. Nonetheless, qualitative conclusions can be drawn by noting [3.55] that the heat conduction flux f_c in a gas is approximately related to the temperature gradient dT/ds by the simple expression.

$$f_c = -\frac{1}{2} n v_{th} \lambda k \frac{dT}{ds}, \qquad (3.33)$$

where v_{th} is the electron thermal speed and λ is the mean free path. For coulomb collisions

$$\lambda_c \propto \frac{v_{th} T^{\frac{3}{2}}}{n},$$

and (3.33) gives a thermal conductivity that is independent of density and proportional to $T^{5/2}$, as in (3.32). If the interactions of the electrons with instability-generated waves occur on a distance scale λ_w smaller than that for collisions, the thermal conductivity will be reduced below that given by (3.32). Further, if λ_w is much less than the scale length of the temperature gradient, the heat conduction flux density will remain proportional to the gradient, with the proportionality constant given by the new, reduced, noncollisional thermal conductivity.

The effects of a modified thermal conductivity on the coronal expansion were first discussed in a general manner by Parker [3.19]. In the context of the approximate one-fluid formulation already described in III.7, Parker demonstrated that a simple change of the constant κ_0 in the conventional plasma conductivity (3.32) was equivalent to changing the coronal density n_0 such that the ratio κ_0/n_0 remained constant. Thermal conductivities proportional to arbitrary powers of the temperature or heliocentric distance still led to supersonic expansion as long as the conductivity was an increasing function of temperature or a decreasing function of distance. In fact, an outward decline in the conductivity generally enhanced the expansion of the corona (the results of subsequent models including the magnetic inhibition of radial heat conduction, as described in III.10, fit into this category and confirm Parker's conclusion). Most relevant to the present discussion is the effect produced by an abrupt cut-off of thermal conduction beyond some heliocentric position r_t. In the region $r > r_t$, the resulting flow is adiabatic, with ultimate conversion of the entire energy flux into kinetic energy. In the class of models considered by Parker, with $T \propto r^{-2/7}$ at large r for the conventional conductivity, the cut-off and subsequent adiabatic flow led to higher expansion speeds.

Specific, empirical forms of a reduced (by both constant and spatially-dependent factors) thermal conductivity have been incorporated into both one-fluid [3.56, 3.57] and two-fluid [3.58, 3.44] coronal expansion models. As $|B_T| \ll 1$ holds deep in the corona, those models that employ the conventional conductivity for r near $r_\odot$ and a reduced value in interplanetary space would appear to be most realistic. For example, Wolff, Brandt, and Southwick [3.44] have incorporated a modified thermal conductivity

$$\kappa = \kappa_0 \, T^{\frac{5}{2}} \left(\frac{r_\odot}{r} \right)^{\tilde{p}},$$

where p is a free parameter, into a two-fluid solar wind model that also includes the magnetic force, magnetic inhibition of conduction transverse to the field, and viscosity. With p taken to produce a fifty-fold reduction in κ between $r = r_\odot$ and $r = 1\,\mathrm{AU}$, the fluid equations were integrated inward toward the sun from an outer boundary of $1000 \, r_\odot$ ($\sim 5\,\mathrm{AU}$). No difficulties with critical points were encountered because

Table 3.13   *A Two-Fluid Solar Wind Model with Reduced Thermal Conductivity*

	$r = 3r_\odot$	$r = 1\,\mathrm{AU}$
Density, cm^{-3}	1.7×10^5	9
Flow Speed, km sec^{-1}	75	303
Proton Temperature, °K	1.3×10^6	4×10^4
Electron Temperature, °K	1.3×10^6	2.0×10^5
Heat Conduction Flux Density, ergs cm^{-2} sec^{-1}		7×10^{-3}

of the presence of the viscous terms (III.11). Table 3.13 summarizes the model judged by the authors to fit both solar wind and coronal data. All of the parameters at 1 AU are close to the observed values. The proton temperature, lower than the electron temperature throughout interplanetary space, became equal to the electron temperature when the integration had preceded inward to $r = 3r_\odot$. The integration was terminated at this position, with the assumption that the basic coronal heating process determines the temperatures (and maintains their equality) for $r < 3r_\odot$. The resulting coronal parameters are reasonable, although the density is somewhat low near the inner boundary. Wolff, Brandt, and Southwick emphasized that this degree of agreement with both coronal and solar wind observations could not be obtained without assuming a modified thermal conductivity. They also concluded that no extended energy source is required to produce a solar wind speed comparable to that observed at 1 AU.

III.14 Another Look at the Energetics of the Coronal Expansion

Our earlier discussion of the energetics of the coronal expansion (III.7), in the context of the basic one-fluid and two-fluid models, produced two fundametally different points of view. The first of these held that an additional, extended source of energy, presumably related to the presence of hydromagnetic waves, must be invoked to produce a realistic description of the expanding corona. The second point of view held that no such energy source is necessary, but that refinement of the basic models, involving in particular the distribution of energy in its various forms, could lead to a realistic coronal expansion model. We are now prepared to re-examine the "energy problem" in the light of the extensions of the basic models described in III.8—III.13.

Unfortunately, such a re-examination leads to few definitive answers. The models described in III.8 demonstrate that the addition of pressure or energy in the region $2r_\odot < r < 20r_\odot$ does produce the expected increase in the expansion speed (and the solar wind proton temperature for a two-fluid model), bringing the predictions of the models into better agreement with observations made near 1 AU. It appears, however, that the properties of solar wind electrons (both the temperature and the heat conduction flux) cannot be adequately explained by this course, even when the coronal electron temperature is assumed to be lower than the coronal proton temperature and the electrons are assumed to be entirely unaffected by the energy dissipation mechanism [3.35]. The models described in III.9—III.13 demonstrate that many other mechanisms can also elevate the expansion speed and raise the proton temperature. The use of a modified thermal conductivity and energy exchange rate largely resolves the problems with the electron temperature and heat conduction flux. It is not entirely clear which, if any, combination of mechanisms can give complete agreement with both coronal and solar wind observations.

In the face of this plethora of (at least partially) proven theories, it becomes clear that we have a relative paucity of observational information regarding the coronal expansion. Solar wind observations have led to detailed knowledge of the solar wind, but in a relatively narrow range of heliocentric distance near 1 AU. Coronal observations give densities in the range $r_\odot \lesssim r \lesssim 20 r_\odot$ along with a gross value of the coronal temperature, but little information regarding the spatial variation of the temperature. Even if plausible questions regarding possible non-spherical aspects of the coronal expansion are put aside (e.g., does the solar wind really emanate from coronal regions that have typical densities), the properties of this complex physical system are known only at two widely separated locations. It would appear that unique conclusions regarding the physical processes that have dominant effects on the coronal expansion can be deduced only by true believers, usually with a parental relationship to one of the competing ideas or models. It should not, in fact, be surprising that the incorporation of several different physical mechanisms can lead to similar modifications of the basic models. Introduction of each new mechanism or assumption is akin to the inclusion of several free parameters in the system of equations, greatly enhancing the possibility of finding a solution that can reasonably connect the observed states at two positions.

To emphasize this basic ambiguity, let us give a final comparison of the two most advanced and throughly described models representing the two different points of view on a problem of energetics: the two-fluid model with an arbitrary extended energy source, as formulated by Hartle and Barnes [3.27] (hereafter referred to as HB), and the two-fluid model including the magnetic force and inhibition of heat conduction, viscosity, and an arbitrary reduction of the electron thermal conductivity, but with no extended energy source, as formulated by Wolff, Brandt, and Southwick [3.44] (hereafter referred to as WBS). Fig. 3.17 shows the number densities predicted by the two models for $2r_\odot \leq r < 20r_\odot$, along with a widely accepted set of observed coronal densities [3.59].

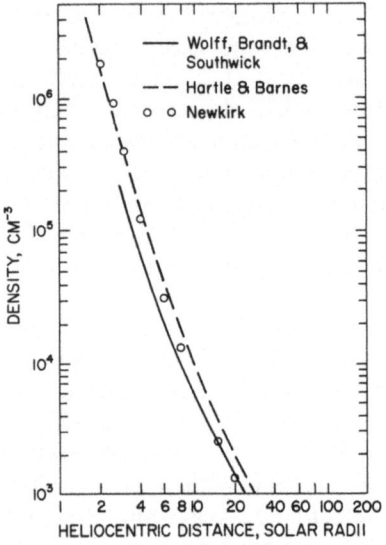

Fig. 3.17   The density $n(r)$ predicted by the two-fluid models of Hartle and Barnes [3.27] (dashed line) and Wolff, Brandt, and Southwick [3.44] (solid line). Observed coronal densities given by Newkirk [3.59] are shown for comparison

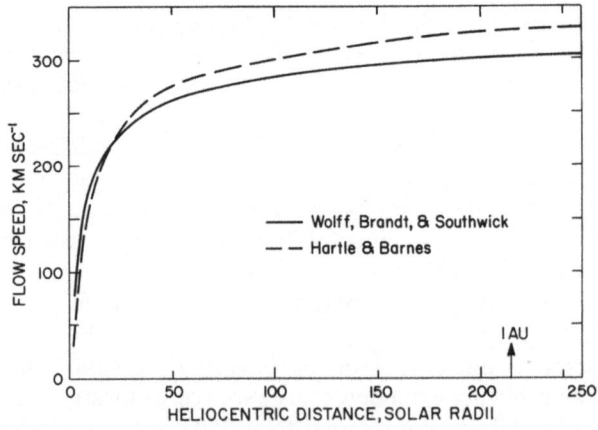

Fig. 3.18   The expansion speed $u(r)$ predicted by the two-fluid models of Hartle and Barnes [3.27] (dashed line) and Wolff, Brandt, and Southwick [3.44] (solid line)

The HB model appears to match the observations somewhat better, especially for $r \lesssim 8r_\odot$, but both models could be said to agree with the observations to within the uncertainties in the latter. The HB model, it should be recalled, predicts a density at 1 AU that is twice as large as observed. Fig. 3.18 displays the predicted radial expansion speeds; there is little difference between the two results. Fig. 3.19 shows the proton and electron temperatures $T_p(r)$ and $T_e(r)$ predicted by the two models. Despite the assumption of a slightly higher coronal electron temperature, the WBS model achieves a lower electron temperature for $r \gtrsim 50r_\odot$ because of the reduction (both by magnetic inhibition and the modification of the classical thermal conductivity) in the electron heat conduction source term. Nearly equal proton temperatures for $r \gtrsim 100r_\odot$ are achieved in the two models by entirely different means. The proton temperature $T_p(r)$ in the HB model is elevated in the range $5r_\odot \lesssim r \lesssim 30r_\odot$ by the extended energy source. The proton temperature $T_p(r)$ in the WBS model is elevated in the range $10r_\odot \lesssim r \lesssim 200r_\odot$ by viscosity. The two profiles are similar only deep in the corona and near 1 AU, where the observational constraints were applied! In summary, both models make some use of the uncertainties in coronal observations (HB in the assumption of different electron and proton temperatures, WBS in the assumption of a somewhat low density). Both give reasonable, though not perfect, agreement with solar wind observations near 1 AU. There appear to be no compelling physical reasons for preferring one model to the other.

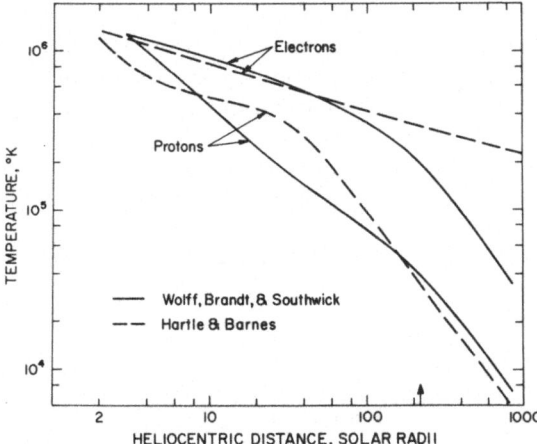

Fig. 3.19   The electron and proton temperatures $T_e(r)$ and $T_p(r)$ predicted by the two-fluid models of Hartle and Barnes [3.27] (dashed lines) and Wolff, Brandt, and Southwick [3.44] (solid lines)

How might this ambiguity be resolved? Minor progress could be achieved if the authors of theoretical models would make direct comparisons with other models, using the *same* coronal conditions, rather than displaying their results for only that set of coronal conditions that places their own model in the best possible light. Further progress might be expected if coronal observations would give more information on the coronal temperature and its spatial variation. It should be mentioned, however, that the great sensitivity of coronal expansion models to the coronal temperature (e. g., III.10) implies that extremely accurate coronal temperatures would be required to serve as an important constraint on the models. The greatest progress can be expected when solar wind observations are performed over a wider range of heliocentric distance. The measurement of coronal expansion speeds at small heliocentric distances by long base-line observations of radio scintillation patterns [3.60] is a promising source of such information. The measurement of solar wind properties (in particular, the temperatures and temperature gradients) between 1 AU and the orbit of Mercury ($r \approx 80 r_\odot$) on spacecraft missions planned in the 1970's offers what is probably the best promise for new information relevant to the problem of the energetics in the coronal expansion.

III.15 Angular Momentum in the Coronal Expansion

One solar wind property given in Table 3.1 remains to be discussed. The average flow direction of the interplanetary plasma is not quite radial from the sun. Rather, the solar wind appears to flow from $1\text{-}\frac{1}{2}°$ east of the sun, as illustrated on Fig. 3.20, with the nonradial velocity component given in the tabulation. In view of the important implications

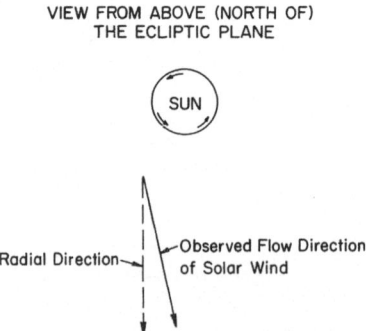

Fig. 3.20   The sense of the observed nonradial flow of the solar wind (viewed from above the solar ecliptic plane) from "east" of the sun

of this result, to be explored presently, it is unfortunate that this is the least certain of all the solar wind properties sumarized in Table 3.1. The 1-$\frac{1}{2}°$ deviation from radial flow is, in fact, only slightly larger than the expected accuracy of the best modern measurements. Differing views of these observations, their limitations, and the credibility of the result can be found in a recent panel discussion of this topic [3.61].

Any nonradial component of the solar wind velocity implies a transport of angular momentum from the sun. For the quiet solar wind conditions given in Table 3.1 the average rates of angular momentum loss from the sun are $n u_r m u_\phi r_e$ (where u_r and u_ϕ are radial and "azimuthal" components of the solar wind speed and r_e is the astronomical unit) $\approx 4.8 \times 10^3$ gm cm^2 sec^{-1} cm^{-2} sec^{-1} for the plasma, and $(B_r B_\phi/4\pi) r_e$ (see below) $\approx 1.5 \times 10^3$ gm cm^2 sec^{-1} cm^{-2} sec^{-1} for the magnetic field (approximate values because $\langle AB \rangle \neq \langle A \rangle \langle B \rangle$). If these losses were uniform over $\pm 30°$ of solar latitude (the observations are made only in the ecliptic plane, or in the range $\pm 7°$ of solar latitude) the implied torque on the sun would be 8.6×10^{30} dyne cm in the sense opposing solar rotation. The angular momentum of the sun, assuming it to be rotating as a solid body, is $\sim 1.7 \times 10^{48}$ dyne cm sec. Thus the torque exerted on the sun by the expanding coronal material is sufficient to brake the solar rotation in $\sim 2 \times 10^{17}$ sec or $\sim 6 \times 10^9$ years. As this braking time is comparable to the generally accepted age of the sun, the loss of solar angular momentum due to the coronal expansion could play a significant role in solar dynamics and evolution.

Coronal material at $r = r_\odot$ near the solar equator has an azimuthal velocity component, u_ϕ, of ~ 2 km sec^{-1}. If there were no nonradial forces on a fluid parcel in the corona and solar wind, u_ϕ would be proportional to $1/r$ (conservation of angular momentum $m u_\phi r$ for the plasma) and the azimuthal velocity component would be only $\sim 10^{-2}$ km sec^{-1} at 1 AU, or about 10^{-3} of the observed average value. Our discussion of refinements of the basic coronal expansion models involved two forces that can have nonradial components, the magnetic and viscous forces, and thus might have an important effect upon the azimuthal velocity component of the solar wind. We will consider here quantitative models of this effect and the comparison of these models with observations.

Under our usual assumptions of spherical symmetry and steady flow, the azimuthal momentum equation in the presence of the magnetic force is

$$mn \frac{u_r}{r} \frac{d}{dr}(r u_\phi) = \frac{1}{c}(j \times B)_\phi$$

$$= \frac{B_r}{4\pi r} \frac{d}{dr}(r B_\phi). \tag{3.34}$$

The constancy of particle flux (3.11) and the $1/r^2$ dependence of B_r (1.24) permit immediate integration of (3.34) to give [3.40]

$$m r u_\phi - \frac{r B_r B_\phi}{4\pi n u_r} = L,\tag{3.35}$$

where L is simply the constant angular momentum of the plasma plus the magnetic field. In the presence of a nonradial velocity component, the magnetic field and velocity components are related by an expression similar to (1.23), namely

$$\frac{B_r}{B_\phi} = \frac{u_r}{u_\phi - \omega r \sin\theta}.\tag{3.36}$$

Use of (3.36) to eliminate B_ϕ from (3.35) gives (at the solar equator where $\sin\theta = 1$),

$$u_\phi(r) = r \frac{\dfrac{4\pi n u_r^2}{B_r^2 r^2 \omega} L - 1}{\dfrac{4\pi m n u_r^2}{B_r^2} - 1}.\tag{3.37}$$

The "radial Alfvén Mach number" $M_A^2 = u_r^2/(B_r^2/4\pi m n)$ is small near the base of the corona but is observed to be ~ 10 near 1 AU. Thus it must pass through unity at some intermediate heliocentric position, r_A. The zero denominator of (3.37) at $r = r_A$ implies that $u_\phi(r)$ is finite and continuous only if the numerator is also zero; r_A plays a role in the azimuthal momentum equation similar to that of the critical radius in the radial momentum equation (I.4). This condition then determines the total angular momentum in the flow from

$$L = m\omega r_A^2.\tag{3.38}$$

$M_A^2/u_r r^2$ is easily shown to be constant, and can be evaluated at $r = r_A$ to give

$$M_A^2 = \frac{u_r r^2}{u_{rA} r_A^2},\tag{3.39}$$

where $u_{rA} = u_r(r_A)$. Substitution of (3.38) and (3.39) into (3.37) finally yields

$$u_\phi = \frac{\omega r}{u_{rA}} \frac{u_{rA} - u_r}{1 - M_A^2}.\tag{3.40}$$

Equation (3.40) has a straightforward physical interpretation. As $r \to \infty$, $M^2 \to (u_\infty/u_{rA})(r^2/r_A^2) \gg 1$, so that

$$u_\phi \to \omega r_A \left(\frac{r_A}{r}\right)\left(1 - \frac{u_{rA}}{u_r}\right).\tag{3.41}$$

If the plasma were to rotate with the angular velocity of the sun out to $r = r_A$, and experience no azimuthal force beyond this radius, one would expect

$$u_\phi \to \omega r_A \left(\frac{r_A}{r} \right)$$

as $r \to \infty$. Equation (3.41) differs from this expression only in the factor $1 - u_{rA}/u_r$, a correction for the angular momentum retained by the magnetic field at large r. In fact, Parker [3.62], in his original discussion of the solar wind, predicted the maximum torque on the sun by means of a similar intuitive argument, that the plasma could be constrained to rigid rotation by the magnetic field only in the region where the magnetic energy density was higher than the energy density in the expanding plasma, or $r \lesssim r_A$.

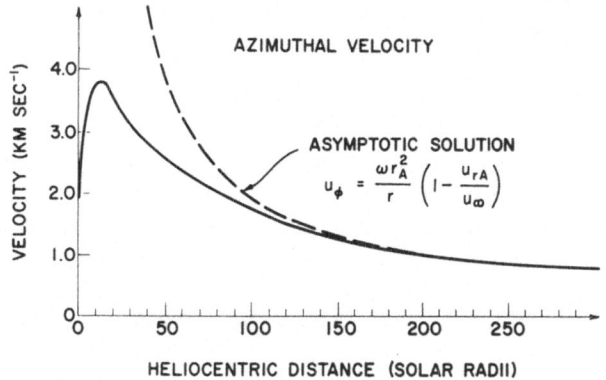

Fig. 3.21   The solar wind azimuthal velocity component produced by the magnetic field in the model of Weber and Davis [3.40]

The analysis given above was first incorporated into quantitative coronal expansion models by Weber and Davis [3.40], Modisette [3.63], and Alonso-Faus [3.64]. The solution $u_r(r)$ derived by Weber and Davis has already been described in III.9. The solution $u_\phi(r)$ is shown in Fig. 3.21. The approach to the limiting form (3.41) is illustrated. It should be emphasized that the coronal plasma does not in actuality rotate rigidly with the sun out to the position $r = r_A$ (at 24.3 $r_\odot$ for this example). The cumulative effect of the nonradial magnetic force over a large range of r, including $r > r_A$, brings $u(r)$ to its limiting value. The azimuthal velocity component predicted at 1 AU by Weber and Davis was only $\sim 1 \; \mathrm{km \; sec^{-1}}$, much lower than the $\sim 8 \; \mathrm{km \; sec^{-1}}$ inferred from spacecraft observations. Most of the angular momentum is carried by the magnetic field at all heliocentric distances in this model.

Brandt, Wolff, and Cassinelli [3.42] considered the nonradial flow produced by the magnetic force in a model including direct integration of the one-fluid energy equation (the Weber and Davis model was based on the assumption of a polytropic expansion). The resulting $u_r(r)$ gave a larger value of $1 - u_{rA}/u_r$, and thus a more efficient transfer of the total angular momentum to the plasma. The azimuthal velocity component predicted at 1 AU was 2.5 km sec^{-1}. The disparity between these two models, differing only in the technique used to solve the energy equation, demonstrates an extreme sensitivity of the solution $u_\phi(r)$ to the form of $u_r(r)$. The incorporation of additional changes into the energy equation (e.g., heating mechanisms or a modified thermal conductivity) could lead to further large changes in $u_\phi(r)$.

Weber and Davis [3.48] were also the first to consider the effect of the viscous force on the nonradial flow of the solar wind. For a steady, spherically-symmetric flow, the conservation of angular momentum (3.35) can be expressed in the general form

$$mru_\phi - \frac{r\,T_{r\phi}}{n\,u_r} = L\,,$$

where $T_{r\phi}$ is the $r\phi$ component of the total stress tensor, and L is the constant total angular momentum. In addition to the magnetic stress, $B_r B_\phi/4\pi$ used above, viscosity adds a component of rather complex form. Other components, such as the stress due to an anisotropic pressure tensor, are easily added to this general formulation. Weber and Davis assumed the same solution $u_r(r)$ from the inviscid model described above,

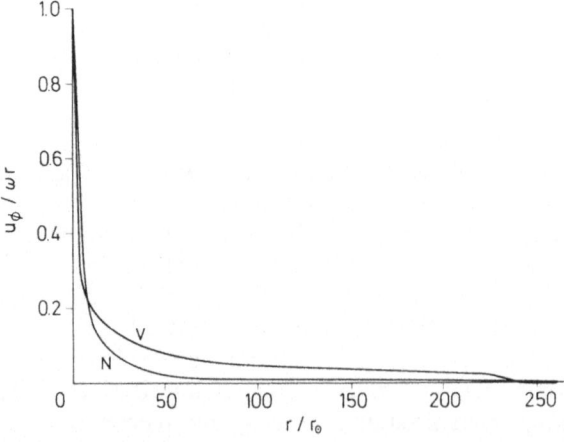

Fig. 3.22 The azimuthal velocity component produced by the magnetic field (N) and the magnetic field, viscosity, and pressure anisotropy (V) in the model of Weber and Davis [3.48]

and an anisotropic pressure tensor wherein the ratio of pressures $P_{||}$ parallel and $P_\perp$ transverse to the magnetic field varied from unity near the sun to 2 (the observed value for solar wind *protons* at 1 AU) at large r. The resulting $u_\phi(r)/\omega r$ is shown as the curve labeled V in Fig. 3.22. The earlier inviscid solution N is also given for comparison. The viscous model predicts a larger u_ϕ in the range $10r_\odot \lesssim r \lesssim 1$ AU, an effect directly attributable to the viscous force. At 1 AU, $u_\phi = 6$ km sec^{-1}, a value much larger than that produced by the magnetic force alone and comparable to that inferred from observations. The reversal in sign just beyond 1 AU is an effect of the pressure anisotropy term.

The Weber and Davis model illustrates the basic effects of the viscous force and thermal anisotropy on the nonradial flow of the solar wind. The detailed applicability of the model can be questioned for two reasons. The one-fluid polytropic expansion basic to the model predicts a temperature of 2×10^5 °K at 1 AU, about four times the proton temperature observed in the quiet solar wind. As the classical plasma viscosity used in the model is proportional to $T^{5/2}$, the viscous force (the source of the larger u_ϕ) has been significantly overestimated. The *total* pressure tensor (III.3) observed near 1 AU is not as anisotropic as assumed in this model (and it is expected that the condition $P_\perp > P_{||}$ will develop beyond 1 AU [3.19]). It would thus also appear that the effect of the pressure anisotropy has also been overestimated.

Wolff, Brandt, and Southwick have included the viscous and magnetic forces in the azimuthal momentum equation for the two-fluid model already discussed extensively in the context of the energy problem (III.11, III.13, and III.14). Actual integration of the two-fluid energy equations again replaces the polytropic law assumption of Weber and Davis, leading to a more realistic proton temperature near 1 AU. As might be expected, the viscous force is reduced and a much lower value of u_ϕ, 1.5 km sec^{-1}, results at 1 AU. The difference between these two models demonstrates an extreme sensitivity of the solution $u_\phi(r)$ to the forms of both $u_r(r)$ and $T_p(r)$.

If any general conclusion can be drawn from the models described above, it would probably be that they predict a nonradial velocity component significantly smaller than that given in Table 3.1. We should recall the possibility that the observed value, $\langle u_\phi \rangle \approx 8$ km sec^{-1}, used in this comparison, is of marginal credibility. It should also be recalled that the coronal expansion models in which nonradial forces and flow have been considered remain highly idealized. Schubert and Coleman [3.65] have pointed out that Alfvén waves (II.3 and III.8) can carry angular momentum from the sun and that this angular momentum might be transferred to the plasma. Coronal and interplanetary inhomogeneities can also lead to azimuthal stresses and motions. The viscosity of the nearly collisionless interplanetary plasma might be significantly different from the classical expression used here and in III.11. In light of the known observational uncertainties and theoretical idealizations, it would seem somewhat premature to regard the differences between the models and observations as a fundamental problem.

III.16 Evaporative Models of the Coronal Expansion

With the advent of *in situ* solar wind observations, the controversy regarding applicability of supersonic or subsonic expansion models to the solar corona was settled in favor of the former. By implication, the use of evaporative models, as employed by Chamberlain to support his arguments for a subsonic expansion (I.6), also appeared to be inappropriate. Such models thus fell into some disfavor.

A revival of interest in evaporative coronal models has since occurred. The harbinger of this revival was provided by Brandt and Cassinelli [3.66], who made two refinements of Chamberlain's model. The first refinement involved Chamberlain's use of a single collision time (that appropriate to particles moving at the mean thermal speed) to define the critical level above which there are no collisions. It is, in fact, the ions with above-average thermal speeds that most readily escape solar gravity. These ions have longer collision times and thus a lower critical level than that of ions with the mean thermal speed. For a corona with $T_\odot = 1.5 \times 10^6 \, °K$, incorporation of a particle energy-dependent critical level in the model raised the expansion speed predicted at 1 AU from $\sim 10 \, km \, sec^{-1}$ to $140 \, km \, sec^{-1}$. The second refinement involved introduction of closed magnetic field lines that "shut off" the escape of particles from some coronal regions. The particles escaping from other regions then collide with a background whose density declines more rapidly with r than if all of the corona were "open." With half of the corona assumed to be "closed" at $r = r_\odot$, the expansion speed predicted at 1 AU was increased to $220 \, km \, sec^{-1}$ for the same $T_\odot = 1.5 \times 10^6 \, °K$ corona. For $T_\odot = 2 \times 10^6 \, °K$, and various coronal density profiles, the combination of these two refinements led to models with the following predicted properties near 1 AU:

Proton Density	2.85 to 3.59 cm^{-3},
Expansion Speed	255 to 278 km sec^{-1},
Proton Temperature	0.87 to $1.5 \times 10^5 \, °K$.

These predictions agreed with solar wind observations about as well as those derived from Parker's fluid models, and led Brandt and Cassinelli to the conclusion that "the solar wind appears to be a natural property of the corona, and does not fundamentally depend on the method of treatment (e. g., hydromagnetic versus exospheric, etc.)."

The most important recent advance in our understanding of evaporative models stems from a reexamination of the electric field produced by polarization of the coronal plasma (I.6). Jockers [3.67] pointed out that both Chamberlain and Brandt and Cassinelli had used the classical Pennekoek-Rosseland electric field, derived under the assumption of *static* equilibrium. This electric field ensures neither charge neutrality nor the absence of a net flow of charge from the sun in an expanding atmosphere. The proper field for the expanding (in the sense of particle evaporation) case must be derived from the momentum conservation (not static equilibrium) equation (see IV.4). Jockers carried out the solution of the coupled equations of motion and charge conservation under several different assumptions. One of these treated both electrons and protons as collisionless above $r_0 = 2.5 r_\odot$, where the density and temperature were assumed to be $n_0 = 9 \times 10^5 \text{ cm}^{-3}$ and $T_0 = 1.32 \times 10^6 \text{ °K}$. The resulting density and expansion speed at 1 AU were 18 cm^{-3} and 172 km sec^{-1}. In three other models the difference between electron and proton collision times was recognized and the electrons treated as collision-dominated (essentially hydrostatic) and only the protons as collisionless above the same $r = 2.5 r_\odot$ base. Three different electron temperature profiles were then *assumed* for $r > 2.5 r_\odot$:

$$
\begin{array}{lll}
\text{Case I} & T_e = 1.32 \times 10^6 \text{ °K} & \text{for } r \leq 9 r_\odot, \\
& T_e \propto r^{-\frac{1}{2}} & \text{for } r > 9 r_\odot; \\
\text{Case II} & T_e = 1.32 \times 10^6 \text{ °K} & \text{for } r \leq 9 r_\odot, \\
& T_e \propto r^{-\frac{1}{3}} & \text{for } r > 9 r_\odot; \\
\text{Case III} & T_e = 1.32 \times 10^6 \text{ °K} & \text{for } r \leq 25 r_\odot, \\
& T_e \propto r^{-\frac{1}{2}} & \text{for } r > 25 r_\odot.
\end{array}
$$

Table 3.14 *Solar Wind Properties at 1 AU Predicted by Evaporative Models of the Coronal Expansion*

	Case I	Case II	Case III
Density, cm^{-3}	14	12	11
Expansion Speed, km sec^{-1}	253	288	322
Proton Temperature, °K	8.1×10^4	6.7×10^4	5.6×10^4

Table 3.14 gives the solar wind properties predicted at 1 AU for the three assumptions. These models clearly demonstrate that evaporative models based on reasonable coronal conditions can lead to expansion

speeds as high as those deduced from the solar wind solutions to the fluid equations.

Hollweg [3.68] constructed evaporative coronal expansion models in which the electrons were treated as collision-dominated and isothermal at all heliocentric distances, while the protons were treated as a fluid below a heliocentric position r_f, but collisionless above r_f. The basic difference between this treatment and that of Jockers is that r_f was taken to be greater than r_c, the critical radius. Starting with fluid parameters predicted at $r=r_f$ by the basic two-fluid model, Hollweg followed the motion of the ions for $r>r_f$ in the presence of solar gravity, a radial magnetic field, and the correct polarization electric field. For $10 r_\odot \leq r_f \leq 20 r_\odot$ and an electron temperature T_e near 10^6 °K, the predicted flow speed near 1 AU was above 300 km sec^{-1} and the proton temperature was $\sim 10^4$ °K. Both of these values are in better agreement with solar wind observations than the predictions of the basic fluid models (Table 3.5). A subsequent treatment by Hollweg [3.69] that included a uniform temperature gradient and a spiral magnetic field led to predictions in poorer accord with solar wind observations.

What is the place of these refinements of evaporative models in our present day concept of the coronal expansion and solar wind? Although these models do predict some solar wind parameters that agree with observations as well as or, in a few cases, better than do fluid models, they fail to agree with observations in several important ways. The most striking disagreement is in the anisotropy of the proton distribution function. In a collisionless expansion with no magnetic field, the spread of radial velocities (equivalent to a "radial temperature") about the mean changes only slightly. However, the spread of nonradial velocity (equivalent to a "transverse temperature") decreases as $1/r$ due to angular momentum conservation for each particle. Thus a highly anisotropic distribution function should develop in interplanetary space. Jockers [3.67] found a ratio K of radial to transverse temperatures between 1100 and 790 at 1 AU for the three models summarized on Table 3.14. In the presence of a radial magnetic field a similar anisotropy is expected, as conservation of the magnetic moment $v_\perp^2/B$ where $v_\perp$ is the nonradial component of a particle velocity, again leads to $v_\perp \propto 1/r$. Thus Hollweg [3.68] deduced an anisotropy ratio $K \approx 50$ (smaller than that of Jockers because the protons in Hollweg's model were taken to become collisionless at a larger heliocentric distance) near 1 AU. Introduction of the spiral magnetic field reduces the anisotropy ratio K to ~ 10 (even less if collisions are assumed to be important to $40 r_\odot$) [3.69]. The observed proton anisotropies near 1 AU give $\langle K \rangle \approx 2$ [3.9], indicating that some interaction process has limited the growth of the anisotropy, keeping the pressure tensor near the isotropy implicitly assumed in the fluid models.

The second disagreement concerns the motion of ions with masses and charges different from that of protons [3.70]. The ion He^{++} is routinely observed in the solar wind, and found to be moving with very nearly the same mean expansion velocity as the protons or H^+ ions (IV.2). One would, however, expect the evaporation speeds of ions with different charge to mass ratios to be quite different. The observed near-equality of $^1H^+$ and $^4H^{++}$ mean velocities in the solar wind indicates that these ions are strongly coupled by some interaction mechanism, and thus are not collisionless. The ability of fluid models to predict solar wind disturbances similar to those actually observed (Chapters V and VI) also implies some such interaction. Despite these indications that solar wind ions do behave as a fluid (i.e., collectively) rather than a collision-free gas, the evaporative models can be used as a starting point in studying such interaction mechanisms (such as plasma instabilities) and their development in the expanding corona [3.71, 3.72].

Finally, these refinements point to a resolution of the still remaining paradox regarding the applicability to the corona of supersonic or subsonic solutions of the fluid equations. Chamberlain's argument (I.6) that the proper fluid model should resemble the evaporative solution appears, in retrospect, to have been valid. However, the evaporative models known at the time of this debate were based on the assumption of an electric field appropriate to a static atmosphere, and thus did not predict the rapid evaporative expansion now shown to be possible when the correct self-consistent electric field is employed.

Chapter IV

Chemical Composition of the Expanding Coronal and Interplanetary Plasma

IV.1 Introduction

The chemical composition of the interplanetary plasma is of interest
not only as a characteristic of the solar wind but as a possible indication
of the chemical composition of the outer layers of the sun. Only two of the
expected chemical constituents of the plasma, the ions $^1H^+$ and $^4H^{++}$
(where, in our notation, the left superscript denotes the atomic mass
number and the right superscript denotes the ionic charge) have been
routinely and extensively observed in the solar wind. These observations
have led to our only firm conclusions regarding an average elemental
abundance in interplanetary space; the solar wind has an average helium
abundance between 4 and 5% by number. Comparison with the solar
helium abundance suggests (but by no means proves) that helium is concentrated in the solar corona as a consequence of the coronal expansion.
Such a separation of different ions could result from the polarization
electric field of the coronal plasma, as already described in I.6. Ions
other than $^1H^+$ and $^4H^{++}$ can be observed by present day plasma
detectors only under favorable (and rare) circumstances. These infrequent
observations have led to some knowledge of the interplanetary abundance
of ^3He, O, Si, and Fe, and the ionization states of O, Si, and Fe. Even
this meager information concerning the ionization states is of great
interest, as it leads to a direct indication of the temperature of the coronal
region from which the observed solar wind has emanated. There are also
some hints that the interplanetary ionization state may be slightly modified by interaction with interstellar gas and dust, raising the possibility
of inferring properties of the interstellar medium from solar wind
observations.

In pursuing the topic of interplanetary chemical composition, we
will come closer to a discussion of basic solar wind data than elsewhere
in this volume. This approach is adopted to convey to the reader both
an understanding of the limitations of present-day observations and an
appreciation of the results obtained within these limitations. Our earlier

discussion of the basic dynamics of the coronal expansion will be generalized to include constituents other than protons and electrons, and the results applied to discussing the relationships among the photospheric, coronal and interplanetary helium abundances. As our understanding of these generalized models and their implications remains rather primitive, the discussion will be carried out entirely within the context of a structureless coronal expansion (i.e., assuming a steady, spherically-symmetric flow of plasma), and is relevant to a broad (in much the same sense as Parker's models described in I.4), rather than detailed, description of the pertinent physical processes.

IV.2 The Determination of the Average Solar Wind Helium Abundance

The basic technique by which most of our information regarding the chemical composition of the solar wind has been acquired is extremely simple. With a single exception, plasma detectors have analyzed the stream on the basis of the energy per charge of its constituents. For of the potential producing this deflection permits separation of the stream on the basis of the energy per charge of its constituents. For example, Fig. 4.1 shows the positive ion flux measured by the Mariner 2 plasma detector (that accepted all particles moving within $\sim 6°$ of the radial direction) at a series of logarithmically spaced energy-per-charge channels. The supersonic nature of the solar wind leads to a narrow

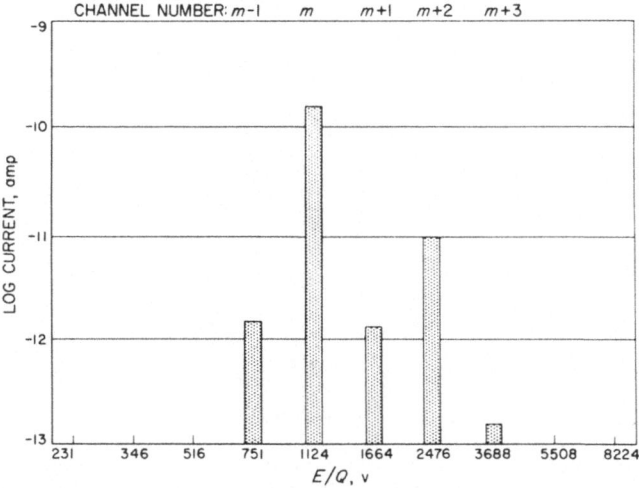

Fig. 4.1 A typical energy-per-charge spectrum for solar wind positive ions observed by the Mariner 2 spacecraft [4.1]

flux peak for any given ionic species. The presence, as in Fig. 4.1, of two distinct flux peaks, the smaller at twice the energy per charge of the larger, was a persistent feature of the Mariner 2 observations. These two peaks were attributed to $^1H^+$ and $^4H^{++}$ ions traveling at a common mean speed; the helium ions would then have twice the energy per charge of the hydrogen ions, leading to the observed positive ion spectrum [4.1]. Note that no discernible flux peak occurs at four times the energy per charge of the primary peak, where any $^4H^+$ ions would be expected (assuming a common expansion speed). Thus little $^4He^+$ would appear to be present.

The existence of similar primary and secondary peaks in energy-per-charge spectra has been reported from virtually every plasma detector operated in the solar wind [4.1, 4.2, 4.3, 4.4, 4.5, 4.6, 4.7, 4.8]. The identification of these peaks as the $^1H^+$ and $^4He^{++}$ components of the plasma has been almost universally accepted, largely, of course, because these would be expected to be the two most abundant ions in coronal material. Support for this identification was provided by a pulse-size analysis of the electron multiplier signals at the two peaks performed on the Vela 3 spacecraft by Bame et al. [4.9]. The most conclusive evidence supporting the conventional identification was obtained by a plasma detector system flown on the Explorer 34 spacecraft by Ogilvie et al. [4.7]. In addition to the usual electrostatic analysis, this instrument subjected the solar wind ion stream to the force produced by perpendicular electric and magnetic fields, leading to an additional analysis of the

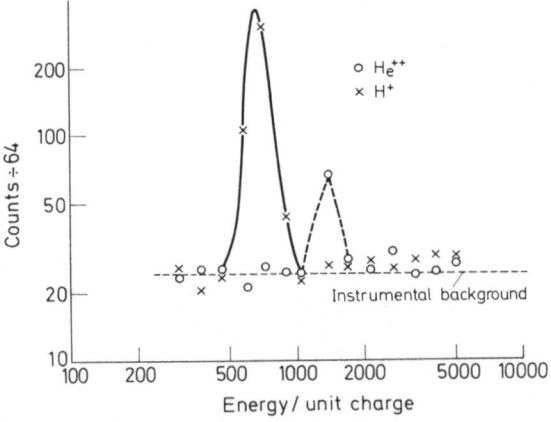

Fig. 4.2   Typical Explorer 34 energy-per-charge spectra at the mass to charge ratios $M/Q=1$ and 2 atomic mass units per electronic charge. The first of these, corresponding to $^1H^+$, is indicated by the x's, while the second, corresponding to $^4He^{++}$, is indicated by the o's [4.7]

speed of the ions. Fig. 4.2 shows energy-per-charge spectra with the speed analysis applied so as to transmit ions with mass-to-charge ratios, M/Q, of 1 and 2 atomic mass units per electronic charge at each channel. In the primary spectral peak, the flux of ions with $M/Q=2$ (e.g., $^4He^{++}$) is at the instrumental background noise level, indicating that most of the ions producing this peak have M/Q near 1, or are $^1H^+$ ions. In the secondary spectral peak, the flux of ions with $M/Q=1$ (or $^1H^+$) is at the instrumental noise level, indicating that most of the ions producing this peak have M/Q near 2, or are probably $^4He^{++}$ (although ions such as $^2H^+$ and completely ionized heavier ions are not unambiguously eliminated).

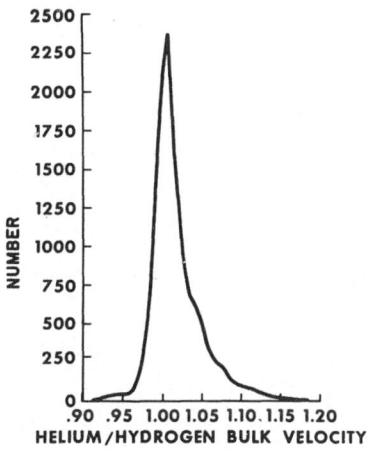

Fig. 4.3   The distribution of the ratios of $^4He^{++}$ and $^1H^+$ flow speeds inferred from Vela 3 energy-per-charge spectra [4.10]

The location of the secondary and primary flux peaks at an energy-per-charge ratio of two (and by implication the equality of $^4He^{++}$ and $^1H^+$ expansion speeds at 1 AU) has been shown to hold to a remarkable degree. Fig. 4.3 shows a distribution of the ratios of the flow speeds inferred from the centroids of the spectral peaks observed by the Vela 3 spacecraft [4.10]. The Vela 3 analyzer systems employ a sufficient number of narrow, closely-spaced energy-per-charge channels to define the mean speed associated with a given spectral peak with an accuracy of ~2%. Fig. 4.3 indicates that the helium and hydrogen flow speeds are equal to this accuracy. Ogilvie and Zwally [4.11] have drawn a similar conclusion from the Explorer 34 observations (with unambiguous separation of the two spectral peaks).

The relative abundance of helium and hydrogen is easily (at least in principle) determined from energy-per-charge spectra such as Figs. 4.1 and 4.2 by comparison of the areas under the secondary and primary spectral peaks. In practice, the imperfect resolution and delineation of these peaks leads to uncertainties in the determination or requires assumptions in performing the comparison. These practical difficulties have become less important as spacecraft instrumentation has become increasingly sophisticated. Four long-term studies of the solar wind helium content have been published and will form the basis of the forthcoming discussion [4.1, 4.8, 4.10, 4.12]. The observed ratio of number densities, $h = n_{He^{++}}/n_{H^+}$ is found to vary widely; Fig. 4.4 shows the

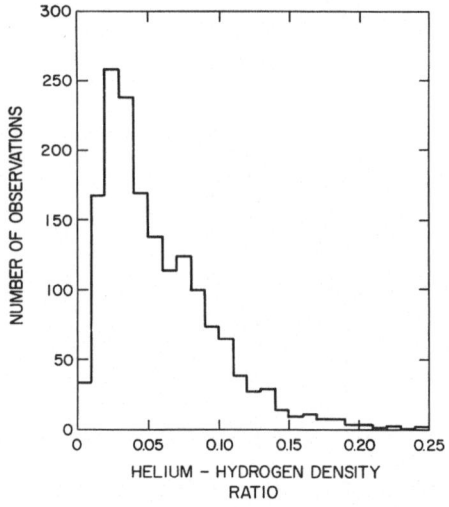

Fig. 4.4 The distribution of the ratios of helium and hydrogen number densities observed by the HEOS-1 satellite [4.8]

distribution of h deduced from solar wind observations made by the HEOS-1 satellite [4.8]. The *average* helium abundance ratio (essentially equal to h because $n_{He^+} \ll n_{He^{++}}$) determined in each of the four independent studies is given in Table 4.1, along with pertinent information regarding the source and extent of the data and assumptions and selections implicit in its analysis. The average abundance ratios all fall between 0.037 and 0.055, indicating a reasonable consistency among the independent observations. An upward trend may be indicated by the small differences among the determinations; Hirshberg [4.13] has shown that these differences largely reflect an increasing frequency of unusually large values of h, probably associated with changes in solar activity (see

Table 4.1 *Observational Determinations of the Average Relative Abundance, h, of Helium and Hydrogen in the Solar Wind*

Source	Time Interval	$\langle h \rangle$	Number of Measurements	Comments
Mariner 2	Aug 29 to Dec 30, 1962	0.046	1,213	Energy-Per-Charge Analysis Only; Assumptions Concerning Temperatures; 10% of Data Included in Analysis; Sample Favors Low Solar Wind Speeds
Vela 3	July, 1965 to July, 1967	0.037	10,314	Energy-Per-Charge Analysis Only; 60% of Data Included in Analysis
Explorer 34	May 30, 1967 to Jan 1, 1968	0.051	2,705	Energy-Per-Charge and Mass Analysis; 5% of Data Included in Analysis; Sample Favors High Solar Wind Density
HEOS-1	Dec, 1968 to March, 1969	0.055	1,632	Energy-Per-Charge Analysis Only. Assumption Concerning Temperature for 81% of Cases

Chapter VI). The helium abundance may also be related to the flow speed [4.14] of observation. A reasonable and apparently well determined value for the average helium content of the solar wind would then appear to be $h = 4.0$ to 4.5% relative (by number) to hydrogen.

A second observational technique that has provided some information regarding helium in the interplanetary plasma involves the trapping of ions incident upon an aluminum foil placed in the solar wind. Subsequent laboratory analysis of the recovered foil yields an accurate determination of the amount of accumulated helium (or other element) [4.15, 4.16, 4.17]. This technique has the advantage of yielding a direct measure of the elemental content; for example, any $^4He^+$ present in interplanetary space would be trapped and analyzed along with the $^4He^{++}$. Its disadvantage is the difficulty of deployment; its use has been confined to foil exposures made during manned lunar landings.

The foil-trapping technique determines only the average (over the time of exposure) helium flux density, since any trapped hydrogen can not be distinguished from residual contaminants in the foil with presently feasible exposure times. Helium flux densities of 6.2 and $8.1 \times 10^6 \, cm^{-2} \, sec^{-1}$ were deduced [4.17] from the foil exposures on the Apollo 11 and

12 lunar landings. The quiet time solar wind proton flux density of Table 3.2 and the average relative helium abundance of 4.5% give an expected average helium flux density of $\sim 1.1 \times 10^7 \, cm^{-2} \, sec^{-1}$. The above values are in reasonable agreement with this expectation, reinforcing our confidence in the basic accuracy of the 4.5% value for the solar wind helium abundance. In the case of the Apollo 12 deployment, Vela 5 observations made several times during the foil exposure revealed a nearly steady proton flux density of $1.9 \times 10^8 \, cm^{-2} \, sec^{-1}$. Combination of the two sources leads to a relative helium abundance of 4.3% in excellent agreement with the long-term average value inferred from the observation based on electrostatic analysis.

IV.3 A Comparison of the Solar and Interplanetary Helium Abundances

The solar origin of the interplanetary plasma suggests that some relationship will exist between the chemical composition of the outer layers of the sun and that of the solar wind. The comparison of observed compositions is a natural first step in the examination of this relationship. Unfortunately, the abundance of helium, the only element other than hydrogen routinely observed in the solar wind, is not easily determined for the sun, and may in fact be poorly known. At the photospheric temperature of $\sim 6000 \,°K$, the conventional spectroscopic determination of abundance based on absorption lines is not possible. Helium emission from the chromosphere and from prominences is observed, but the interpretation of these observations is complicated by possible deviations from thermodynamic equilibrium and inhomogeneities in such regions. Unsold [4.18] concluded that intensities of hydrogen and helium emission lines were not inconsistent with a helium-hydrogen abundance ratio of 16%. Hirayama [4.19] has recently refined the study of emission line intensities to derive a helium abundance of $6.5 \pm 1.5 \%$ for the chromosphere.

The general solar helium abundance has also been deduced by three indirect methods:

1. Models of stellar interiors predict a mass-luminosity relationship that depends upon the helium content of the star. Application of these models to the sun has given helium-hydrogen abundance ratios (by number) of 0.095 [4.20, 4.21], 0.0865 [4.22], and 0.077 to 0.087 [4.23]. All of these values pertain to the solar interior and need not apply to the outer layers of the sun.

2. The abundance of helium relative to carbon, nitrogen, and oxygen in solar cosmic rays has been found to be relatively constant [4.24]. Combination of the cosmic ray abundances of helium relative to these

ions with spectroscopic values of the solar abundances of carbon, nitrogen, and oxygen relative to hydrogen gives the relative abundance of helium and hydrogen as 0.09 [4.24, 4.25], 0.063 ± 0.015 [4.26] and 0.062 ± 0.008 [4.27]. This method is based on the assumption that no separation of helium and hydrogen occurs in the solar cosmic-ray energization process.

3. The helium abundance of the sun can be directly related to the flux of solar neutrinos. Iben [4.28] has used the upper limit on the solar neutrino flux [4.29] to infer an upper limit of 0.049 to 0.064 for the helium-hydrogen abundance ratio of the sun. This value again pertains to the solar interior.

Most of these values of the solar helium abundance are larger than the 0.045 value deduced in IV.2 for the solar wind. The helium abundances derived from the cosmic ray-spectroscopic comparisons and from chromospheric emission lines are most directly applicable to the outer layers of the sun; these values are $\sim 50\%$ higher than that deduced for the solar wind. If the solar determinations are taken at face value, they lead to the conclusion that helium is less abundant in the interplanetary plasma than in the solar atmosphere. However, the differences among the solar values determined by different techniques may indicate that experimental and interpretative uncertainties remain comparable to any solar-interplanetary difference that might be inferred.

IV.4 Theoretical Models of the Expansion of a Corona Containing Helium

Our previous discussions of theoretical models of the coronal expansion (Chapters I and III) were based on the assumption that the coronal and interplanetary plasma contained only protons and electrons. In the context of our present interest in the chemical composition of the expanding plasma, we will generalize the theory to a plasma containing additional ions. For simplicity, only the case of a corona composed of fully ionized hydrogen and helium, or with the charged particle constituents $^1H^+$, $^4He^{++}$, and electrons will be presented.

Consider a steady, radial, spherically-symmetric motion of coronal material. The mass conservation equations for each constituent (analogous to equation 3.1) have first integrals that state the constancy of particle flux through any sun-centered spherical surface (as in equation 3.11). The momentum equations for the three constituents are:

$$n_p m_p u_p \frac{du_p}{dr} = -\frac{d}{dr}(n_p k T_p) - \frac{n_p m_p G M_\odot}{r^2} + n_p q E + C_p, \qquad (4.1)$$

$$4 n_\alpha m_p u_\alpha \frac{du_\alpha}{dr} = - \frac{d}{dr}(n_\alpha k T_\alpha) - \frac{4 n_\alpha m_p G M_\odot}{r^2} + 2 n_\alpha q E + C_\alpha, \quad (4.2)$$

$$n_e m_e \frac{du_e}{dr} = - \frac{d}{dr}(n_e k T_e) - \frac{n_e m_e G M_\odot}{r^2} - n_e q E + C_e. \quad (4.3)$$

where the subscripts p, α, and e refer to $^1H^+$, $^4He^{++}$, and electrons, and q is the magnitude of the electronic charge. The terms C_p, C_α, and C_e represent the momentum transfer among the particle species by collisions. All forces other than those due to the pressure and solar gravity have been neglected. The electric field E is that produced by polarization of the plasma, as described in I.6. An energy conservation equation should be also written for each constituent. The integration of this set of mass, momentum, and energy conservation equations is a still more formidable task than that already described in detail in Chapter III. As a reasonable first simplification, let us eliminate the energy equations by assuming a known temperature or a polytropic law. Three levels of approximation have been employed in solving the resulting system of equations.

A first, very crude, approximation, which nonetheless reveals the basic physical properties of the solutions, is that of a static corona [4.30]. Assume that helium is everywhere sufficiently less abundant than hydrogen as to produce only a small perturbation of the conditions in a proton-electron atmosphere. For an isothermal corona with $T_p = T_e = T_\alpha = T$, hydrostatic equilibrium of the protons and electrons gives

$$0 = - k T \frac{dn_p}{dr} - \frac{n_p m_p G M_\odot}{r^2} + n_p q E \quad (4.4)$$

and

$$0 = - k T \frac{dn_e}{dr} - \frac{n_e m_e G M_\odot}{r^2} - n_e q E, \quad (4.5)$$

as the momentum exchange terms C_p and C_e must be zero if there is no plasma flow. Equations (4.4) and (4.5) are identical to (1.29) and (1.30) of I.6, and, under the requirement of charge neutrality, imply the electric field

$$E \approx \frac{m_p}{2q} \frac{G M_\odot}{r^2}. \quad (4.6)$$

Substitution of (4.6) into (4.4) gives

$$k T \frac{dn_p}{dr} = - \frac{1}{2} \frac{n_p m_p G M_\odot}{r^2}, \quad (4.7)$$

which has the solution

$$n_p(r) = n_{p0} \exp \left\{ - \frac{G M_\odot m_p}{2 k T r_0} \frac{r - r_0}{r} \right\}. \quad (4.8)$$

Substitution of the same electric field (4.6) into the hydrostatic equilibrium equation for the $^4\text{He}^{++}$ ions (equation 4.2 with $u_\alpha = 0$ and $C_\alpha = 0$) gives

$$kT\frac{dn_\alpha}{dr} = -3\frac{n_\alpha m_p GM_\odot}{r^2},\qquad(4.9)$$

which has the solution

$$n_\alpha(r) = n_{\alpha 0}\exp\left\{-\frac{3GM_\odot m_p}{kTr_0}\frac{r-r_0}{r}\right\}.\qquad(4.10)$$

Thus the helium density decreases more rapidly with r than does the hydrogen density (the scale heights differ by a factor of 6), giving a decreasing helium abundance with increasing solar altitude.

More general solutions (with non-negligible helium charge density and a variable temperature) for the static corona are given by Parker [4.30]. However, the simple example given above illustrates the basic interplay of the polarization electric field and solar gravity in producing a stratification of ions with different charge-to-mass ratios. For this particular case, (4.7) and (4.9) show that the electric field has canceled half of the gravitational force on an $^1\text{H}^+$ ion, but only one-fourth of the (four times larger) gravitational force on a $^4\text{He}^{++}$ ion. The lighter hydrogen ions thus experience a combined electrostatic and gravitational force only 1/6 that of the helium ions, and essentially "float" at the top of the atmosphere.

The next level of approximation considers the motion of the corona but neglects the momentum exchange terms, again setting $C_p = C_\alpha = C_e = 0$. This approximation has been discussed by Yeh [4.31] with the additional assumption of constant (but not necessarily equal) temperatures for each constituent. The requirements of charge neutrality and no steady electric current (to avoid building up a charge on the sun) give

$$qn_p + 2qn_\alpha - qn_e = 0,\qquad(4.11)$$

and

$$qn_p u_p + 2qn_\alpha u_\alpha - qn_e u_e = 0.\qquad(4.12)$$

Yeh demonstrated that the system of equations (4.1) to (4.3), (4.11), and (4.12) has a critical solution in which all three constituents move subsonically at small r, passing continuously to supersonic flow for the ions at large r. Table 4.2 summarizes one of these solutions for the relative temperatures $T_\alpha = 4T_p$ and $T_e = 2T_p$, in reasonable agreement with observations at 1 AU (but probably not typical of conditions in the corona). The constant proton temperature was taken to be $T_p = 3.52 \times 10^6$ °K (compare with Fig. 1.2 displaying Parker's isothermal solutions). From the point of view of the present discussion, the most important result of

Table 4.2   *Solar Wind Properties at 1 AU by the "Three-Fluid"* ($^1H^+$, $^4He^{++}$, *and electron*) *Model of Yeh with 10% Helium at* $r=r_\odot$

Hydrogen Flow Speed	1363 km sec^{-1}
Helium Flow Speed	1054 km sec^{-1}
Electron Flow Speed	1337 km sec^{-1}
Helium-Hydrogen Density Ratio	0.452

Yeh's model is that $h(r)=n_\alpha(r)/n_p(r)$, assumed to be 0.1 at $r=r_\odot$, decreases to 0.0452 at $r=1$ AU. This change in helium abundance with heliocentric distance stems from the same combination of electrostatic and gravitational forces as in the static model described above. Note, however that the expansion speeds of the $^4He^{++}$ and $^1H^+$ ions at 1 AU differ by 25% for this example. This prediction is at variance with the observed equality of expansion speeds, demonstrated in Fig. 4.3.

The third level of approximation with which the problem has been treated includes the momentum exchange terms C_p and C_α appropriate to coulomb collisions between the $^4He^{++}$ and $^1H^+$ ions [4.32, 4.33]:

$$C_p=-C_\alpha=\frac{4\pi q^4 \ln \Lambda}{m_p}\ \frac{n}{kT}\ \frac{5}{4+T_\alpha}\ G\left(\left[\frac{4}{4+\frac{T_\alpha}{T_p}}\ \frac{m_p}{2kT}\right][u_\alpha-u_p]\right),$$

where

$$G(x)=\frac{\phi(x)-x\,\phi'(x)}{2x^2},$$

and where $G(x)$ is the error function. The resulting system has been solved by Geiss *et al.* [4.33] by reverting to the simplification of a $^4He^{++}$ density everywhere small enough that the charge neutrality requirement involves only the protons and electrons and the collision terms are negligible in the proton and electron momentum equations (4.1) and (4.3). Thus the charge neutrality and zero current requirements are simply $n_e\approx n_p$ and $u_e\approx u_p$. As in the static case, subtraction of the proton and electron momentum conservation equations gives the electric field;

$$E=\frac{m_p}{2q}\left(\frac{GM_\odot}{r^2}+u_p\frac{du_p}{dr}\right). \tag{4.13}$$

The term $u_p(du_p/dr)$ represents the change in the polarization field associated with the expanding, rather than static, nature of the model. This term is positive, so that the field is strengthened by the expansion. Addition of (4.1) and (4.3) eliminates E and gives the familiar momentum equation

$$nmu_p\frac{du_p}{dr}=-\frac{d}{dr}(nk[T_e+T_p])-\frac{nmGM_\odot}{r^2}. \tag{4.14}$$

Geiss *et al.* [4.33] assumed a polytropic law and used the resulting solutions of (4.14) to determine the electric field (4.13). Insertion of this field into the helium momentum equation and assumption of the same polytropic law $(T_\alpha = T_p)$ then reduced the problem of the helium motion to solution of equation (4.2), a standard momentum equation with an added body force $2n_\alpha q E$ and collision term C_α. Solar wind solutions, passing through a critical point, were found by numerical integration. Because of the collision term C_α, the helium expansion speed $u(r)$ depends upon the proton flux $I_p = 4\pi n_p u_p r^2$.

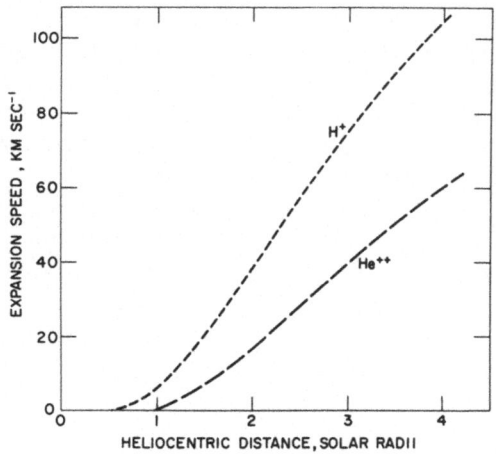

Fig. 4.5   The expansion speeds $u_\alpha(r)$ of He^{++} and $u_p(r)$ of H$^+$ predicted by the coronal expansion model of Geiss *et al.* [4.33]

Fig. 4.5 shows one of the Geiss *et al.* solutions for the ^{4}He^{++} and ^{1}H$^+$ expansion speeds in the corona with a polytropic index $\alpha = 1.1$ and $T_p(r_\odot) = 10^6\,^\circ$K. The proton flux for the ^{4}He^{++} solution is 8.25×10^{35} sec^{-1} (corresponding to a reasonable proton flux density of 3×10^8 cm^{-2} sec^{-1} at 1 AU). The helium ions, again subject to a smaller outward electrostatic force, are accelerated more slowly than the hydrogen. For lower proton fluxes, the difference in the two expansion speeds is greater. At large heliocentric distances $\alpha = 1.1$ leads to a helium flow speed much lower than the hydrogen flow speed. However, the collision term is very sensitive to the temperature and hence to α. Selection of $\alpha > 1.33$ at large r (or a proton temperature falling faster than $r^{-2/3}$) causes the helium flow speed to approach the hydrogen flow speed as $r \to \infty$. For example, $\alpha = 1.4$ $(T \propto r^{-0.8})$ leads to a difference of only 20% in the expansion speeds at 1 AU. Geiss *et al.* suggest that a solution using α

near 1 for small r but $\alpha \geq 1.33$ at large r would be a reasonable representation of the coronal expansion and would thus lead to nearly equal expansion speeds, in basic agreement with observations, at 1 AU.

Constancy of the helium and hydrogen fluxes $n_\alpha u_\alpha r^2$ and $n_p u_p r^2$ leads to

$$h(r) = \frac{n_\alpha(r)}{n_p(r)} = \frac{u_p(r)}{u_\alpha(r)} h(r_0) \tag{4.15}$$

where r_0 is some reference position. Fig. 4.5 predicts a ratio $u_p(r)/u_\alpha(r)$ that decreases with r, implying through (4.15) that h decreases with r in the range $r_0 \leq r \leq 4r_\odot$. If we assume that the helium and hydrogen expansion speeds do approach equality at 1 AU and use $h(r_0) = 0.05$ in interplanetary space, $h(r)$ varies with heliocentric distance as shown in Fig. 4.6. Helium abundance ratios greater than 0.10 result for $r \lesssim 3r_\odot$, with very large ratios near $r = r_\odot$. In fact, the assumption of small $^4\text{He}^{++}$ charge densities breaks down near $r = r_\odot$. Nonetheless, these solutions demonstrate that a separation of ions with different charge-to-mass ratios occurs in the coronal expansion even when the ions are coupled by coulomb collisions. As in the previous approximations, the helium abundance is found to be a decreasing function of heliocentric distance.

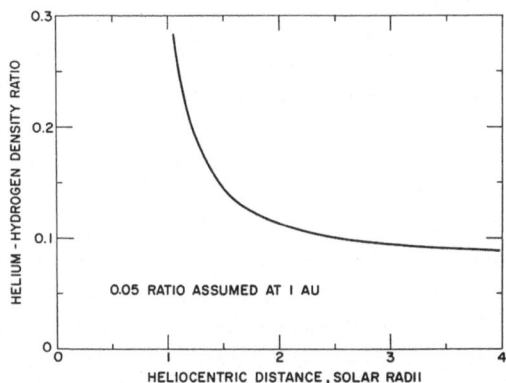

Fig. 4.6   The relative abundance of helium as a function of heliocentric distance implied by the expansion speeds of Fig. 4.5 and the assumption that the helium and hydrogen flow speeds reach equality at 1 AU, where the abundance ratio is observed to be near 0.05 [4.33, 4.34]

It is not entirely clear that the coulomb collision term used in the model of Geiss et al. is adequate to explain the very close approach to equality of helium and hydrogen flow speeds actually observed in the solar wind. Further, the helium abundance ratio has not been found to vary with the proton flux [4.14]. Thus non-collisional processes, similar to those considered in the discussion of energy exchange between electrons and protons (III.12), may be surmised to contribute to the momentum exchange between ^{4}He^{++} and ^{1}H$^+$ ions, leading to an interaction term larger than the C used by Geiss *et al.* The decrease in helium abundance with heliocentric position, predicted in all of the approximations described above, would then occur only if such interactions were unimportant in the lower corona. The relationship between the chemical composition in the corona and solar wind (and the apparent ability of the models described above to reconcile the observed solar wind and solar helium abundances) hinges on such unknown physical processes.

IV.5 The Relationship between the Photospheric and Coronal Helium Abundances

The theoretical models described in IV.4 apply specifically to the corona and interplanetary space; i.e., above the position of the maximum coronal temperature. They have no applicability to layers of the solar atmosphere beneath the temperature maximum—the solar chromosphere and transition zone wherein the temperature changes from the photospheric value of $\sim 6000\,^\circ$K to the coronal value of $\sim 10^6\,^\circ$K. The latter regions differ from the corona in two ways fundamental to the present discussion of chemical composition. The plasma motions are expected to be highly subsonic, and a very steep *inward* temperature gradient must be present. These two conditions imply that diffusive processes will play a dominant role in determining the atmospheric structure and chemical composition. Any attempt to relate the coronal and interplanetary chemical composition, as deduced from the models of IV.4, to the photospheric chemical composition requires examination of diffusion in the chromosphere and transition zone

This problem has been discussed by several authors. Jokipii [4.35] has derived numerical solutions to the hydrogen and helium mass conservation equations including diffusion in an assumed temperature profile. All magnetic effects, inhomogeneities, turbulence, and changes in the ionization state were neglected. The resulting helium abundance ratio $h(r)$ increased with r through the transition zone, reached a maximum value in the lower corona, and then decreased with increasing heliocentric distance. Jokipii suggested that turbulent mixing in the chromosphere and transition zone might remove the compositional differences between the photosphere and low corona, producing a monotonic decrease in $h(r)$ from an assumed photospheric value of 0.1

to the observed interplanetary value of 0.05. DeLache [4.36] has considered the diffusion of small quantities of several ions in a transition zone whose structure was determined by the dynamics of the protons and electrons. The elemental abundance ratios (relative to hydrogen) were again found to increase with altitude. Nakada [4.37] has discussed the diffusion of several ions in a static atmosphere above the level where $T = 10^5\,°K$ (i.e., in the transition zone). With interactions assumed to occur by coulomb collisions, multi-component diffusion equations were written for arbitrary (not necessarily small) concentrations of these ions. Numerical integration led to the by-now-familiar result—an increase in the relative abundances of all elements heavier than hydrogen.

The details of these diffusion models are beyond the scope of this discussion. All predict (in the absence of turbulent mixing) a similar increase in helium or heavier element abundances in the range of solar altitude between the photosphere and the lower corona. The physical meaning of this result is clear and was, in fact, deduced from general principles for a static atmosphere [4.38]. In the presence of the large inward temperature gradient, the lighter hydrogen ions diffuse away from the hot region more rapidly than the heavier ions, producing an equilibrium state in which the heavier ions are concentrated in the hotter region of the atmosphere. However, it is not obvious that such idealized diffusion models can give a quantitative description of the actual solar atmosphere. The solar chromosphere and transition zone are well known to be inhomogeneous and permeated by a complex magnetic field. The magnetic features appear to have a strong influence on the temperature structure of these atmospheric layers. One would then also expect the fields to have a strong influence on the particle diffusion processes pertinent to the models. Mass motions are also observed in the chromosphere, suggesting that turbulent mixing of these layers may be important. In view of these complexities, the conclusions of the simple diffusion models must be applied to the actual solar atmosphere with great care.

IV.6 General Conclusions Regarding the Chemical Composition of the Expanding Solar Atmosphere

If the predictions of the two classes of theoretical models described in IV.4 and IV.5 are combined, a general picture of the helium abundance in the outer layers of the sun emerges [4.39], as illustrated in Fig. 4.7. Helium accumulates in the corona due to the preferential escape of hydrogen (an effect of the combined electrostatic and gravitational forces) in the coronal expansion. As the diffusion of helium back toward the photosphere is less efficient than the diffusion of hydrogen, a steady, equilibrium state must have a higher concentration of helium in the corona than in the photosphere.

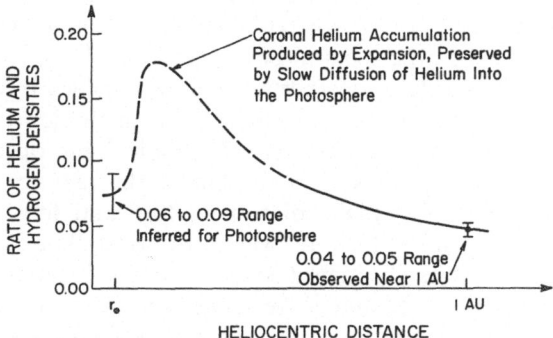

Fig. 4.7 The relative helium abundance in the outer layers of the sun, as implied by the models described in IV.4 and IV.5 [4.39]

Although this picture is plausible, its validity remains largely unproven. The tendency for helium to accumulate in the lower layers of an expanding corona could be defeated by any strong, noncollisional momentum exchange between coronal helium and hydrogen ions. The diffusion equilibrium state would be drastically modified in the presence of turbulent mixing. As the helium abundance has been accurately determined only in interplanetary space and in the chromosphere, present day observations provide no direct test of the picture. One piece of indirect evidence can be mustered: the observation of extremely high helium abundances in solar wind disturbances related to solar flares (Chapter VI) strongly suggests the helium has been concentrated somewhere in the outer solar atmosphere. Despite this indirect evidence supporting our general conclusion, the relationship among the photospheric, coronal, and interplanetary elemental abundances remains an interesting and important unsettled problem.

IV.7 Observations of Solar Wind Ions Other than $^1H^+$ and $^4H^{++}$

There is considerable observational evidence (e.g., Fig. 4.3) that the $^1H^+$ and $^4He^{++}$ ions in the interplanetary plasma are constrained to move with very nearly equal mean speeds. If other positive ions are also assumed to share this common expansion speed, the presence of an ionic species with mass $M m_p$ and charge $Q q$ (where m_p is the proton mass and q the magnitude of the electronic charge) would result in a flux peak in solar wind spectra at M/Q times the energy per charge of the primary $^1H^+$ spectral peak. The detection of any such ions would be difficult for two reasons. The fluxes of all ions other than $^1H^+$ and $^4He^{++}$ are expected to be low due to the relative rarity of other elements and isotopes and due to the nearly complete ionization of ^4He at coronal temperatures. Further, the spectral peaks produced by ions with similar values of M/Q will lie close together and can be resolved by electrostatic analysis only if all such peaks are narrow; i.e., if the ion temperatures are sufficiently low. Because of these difficulties, few observations of such solar wind ions have been made. Under the rare combination of favorable circumstances—high solar wind flux and low ion temperatures—the singly ionized form of ^4He and ions of ^3He, ^{16}O, ^{28}Si, and ^{56}Fe have been tentatively identified in solar wind energy-per-charge spectra. The isotopes ^3He and ^{20}Ne have been found trapped in foils exposed to the solar wind. We will review here these observations and their implications regarding the elemental and ionic composition of the interplanetary plasma.

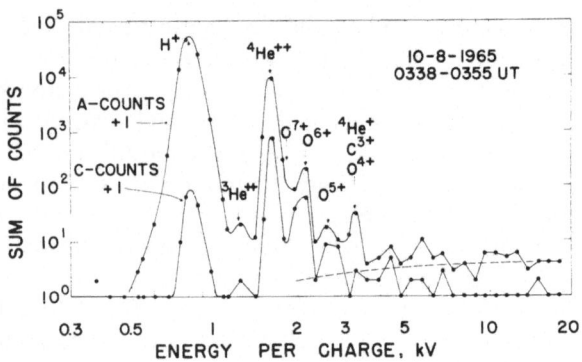

Fig. 4.8   Energy-per-charge spectra observed at two different pulse counting thresholds by a Vela 3 spacecraft [4.9]. The identification of the ionic species producing the various spectral peaks is discussed in the text

Fig. 4.8 shows a set of Vela 3 positive ion data from which the first identification of solar wind ions other than $^1H^+$ and $^4He^{++}$ was established [4.9]. Two spectra are shown, labeled "A-counts" and "C-counts." These represent the numbers of electron multiplier pulses accumulated during the same counting intervals, but with two different pulse-size thresholds. Only very large pulses appear in the "C-counts" spectrum, while nearly all pulses appear in the "A-counts" spectrum. A comparison of the two counting levels at a given energy-per-charge channel gives a crude pulse-size analysis that will prove useful as supplementary information in identifying possible ionic species. The primary spectral peak produced by $^1H^+$ appears at ~0.8 kV in Fig. 4.8. This peak in the C-counts spectrum is only ~10^{-3} as high as that in the A-counts spectrum, showing that $^1H^+$ ions produce very few pulses large enough to exceed the C-threshold. A similar ratio is observed for the primary spectral peak when changing solar wind conditions shift it to other energy-per-charge channels. The secondary spectral peak produced by $^4He^{++}$ occurs, as expected, near 1.6 kV. The C-count level at this peak is ~10^{-1} as high as the A-count level. The $^4He^{++}$ ions at a given energy per charge have a higher energy than $^1H^+$ ions, thus producing larger pulses in the electron multiplier and more C-counts. Other distinct spectral peaks are discernible in Fig. 4.8; the M/Q determined by their location and the rough indication of M given by the ratio of A- and C-counts has led to proposal of the following identifications.

1. The small spectral peak at ~1.2 kV implies an ion with $M/Q = 3/2$. This can only correspond to $^3H^{++}$. The ratio of A- and C-counts is ~10^{-1}, consistent with an ionic mass similar to that at the $^4He^{++}$ peak. Bame et al. [4.9] estimated the abundance of ^3He to be 10^{-3} that of ^4He at the time of this observation. The resolution of the $^3He^{++}$ spectral peak clearly requires that the neighboring $^1H^+$ and $^4He^{++}$ peaks be very narrow (or that the temperature be very low).

2. A prominent spectral peak occurs near 2.1 kV, implying the presence of an ionic species with M/Q near 2.6. The ratio of A- and C-counts at this peak is only ~0.3, indicating the presence of ions of an element heavier than helium (eliminating the possibility that some anomaly of the $^4He^{++}$ distribution function has produced the smaller spectral peak). The most abundant such elements are ^{16}O, ^{12}C, ^{20}Ne, and ^{14}N, so that the ions $^{16}O^{+6}$, $^{12}C^{+5}$, $^{14}N^{+5}$, and $^{20}Ne^{+8}$ are possible sources of the $M/Q = 2.6$ peak. The location of this spectral feature (it is the most commonly observed peak other than those due to $^1H^+$ and $^4He^{++}$) under varying solar wind conditions appears to be most consistent with identification as $^{16}O^{+6}$, and Bame et al. [4.9] have thus proposed that this ionic species is the third most abundant in the solar wind. This result is consistent with (and perhaps partially motivated by) the measured solar abundances of the elements O, C, N, and Ne. Note that no spectral peak is discernible at the energy-per-charge position of $^{16}O^{+7}$, while a very small peak is present at the position of $^{16}O^{+5}$. Thus most of the solar wind oxygen has been ionized to charge $+6q$.

The interplanetary abundance of ^{16}O deduced from the data of Fig. 4.8 is 0.04 that of ^4He. An hour later a value of 0.015 was observed. Similar values have been measured at other times, indicating a substantial variability in the solar wind oxygen abundance. The ionization state of oxygen is generally found to be similar to that in Fig. 4.8, with $^{16}O^{+5}$ (if present at all) much less abundant than $^{16}O^{+6}$, and with $^{16}O^{+6}$ usually more abundant than $^{16}O^{+7}$. This is, in fact, the oxygen ionization state expected at a coronal temperature of 1 to $2 \times 10^6\,°K$, rather than that expected at the observed interplanetary electron temperature of $\sim 10^5\,°K$. The relationship of the "ionization state temperature" to coronal and interplanetary conditions will be discussed in IV.8. The ion $^{16}O^{+7}$ has occasionally been observed to be more abundant than $^{16}O^{+6}$, implying a still higher "ionization temperature".

3. Another distinct spectral peak occurs at $\sim 3.2\,kV$, corresponding to M/Q near 4. A number of ions, such as $^4He^+$, $^{16}O^{+4}$, or $^{12}C^{+3}$ could contribute to this flux peak. However, the A- to C-count ratio is $\sim 10^{-1}$, similar to that at the $^4He^{++}$ peak but not consistent with that at the $M/Q = 2.6$ peak, tentatively identified with the heavier ions of oxygen. This suggests [4.9] attribution of the $M/Q = 4$ peak to $^4He^+$, with the ratio of $^4He^+$ to $^4He^{++}$ densities near 3×10^{-3}. The absence of the $M/Q = 4$ feature at other times implies a lower ratio and again illustrates the variability of solar wind ionic abundances. Alternate identifications of the ionic species producing this spectral feature can be proposed [4.40, 4.41]; for a discussion of these possibilities see the review by Bame [4.42].

Some small spectral features might be discernible beyond the $M/Q = 4$ peak, but the fluxes are only slightly above the average instrumental noise background (indicated by the dashed line on Fig. 4.8), and the

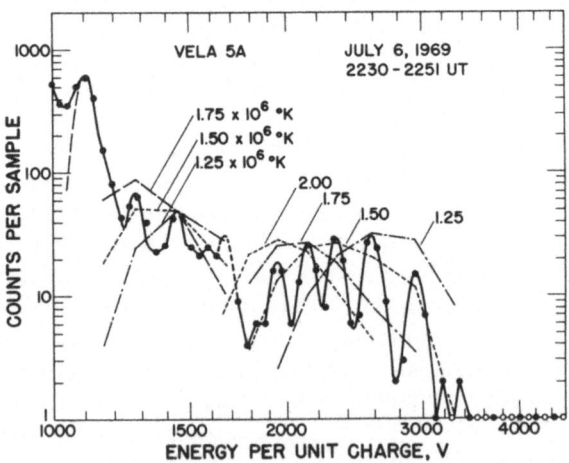

Fig. 4.9   An energy-per-charge spectrum observed on a Vela 5 spacecraft [4.43]. The identification of the ionic species producing the various spectral peaks is discussed in the text

reality of any features in this range of energy per charge is questionable. Subsequent observations by an electrostatic analyzer system flown on the Vela 5 spacecraft have employed higher resolution (more closely spaced energy-per-charge channels) and higher sensitivity to detect several additional spectral peaks at energy-per-charge positions above 1 kV. Fig. 4.9 shows data obtained by this detector system at a time when the primary ($^1H^+$) and secondary ($^4He^{++}$) flux peaks (observed by a second analyzer system covering a lower energy-per-charge range) were located at 445 and 890 volts [4.43]. At least eight additional flux peaks are discernible in these data. Identifications of the ionic species responsible for these features have been proposed [4.43] under the familiar assumption that all ions have a common expansion speed. The spectral peak at 1180 volts then corresponds to $M/Q = 2.65$, and is the same feature assigned to the ion $^{16}O^{+6}$ in the Vela 3 observations described above. The identification of the remaining peaks is based on the implied M/Q values and the expected abundances of the elements in solar material.

4. The two very sharp spectral peaks at 1260 and 1430 kV may be produced by the ions $^{28}Si^{+9}$ and $^{28}Si^{+8}$ (although the peaks produced by sulfur ions would lie nearby and may confuse this identification). The expected envelope of the silicon ionization state at several coronal temperatures is indicated on Fig. 4.9; the relative heights of the two peaks tentatively attributed to silicon are consistent with an "ionization temperature" between 1.5 and $1.75 \times 10^6\,^\circ K$. A third spectral peak would then be expected at ~ 1700 kV. No such peak is discernible, but an ill-placed gap in the energy-per-charge channels might account for its absence.

5. The five flux peaks in the range 1800 to 3000 volts occur at the energy-per-charge locations that correspond to the ions $^{56}Fe^{+13}$ to $^{56}Fe^{+8}$. These identifications are highly plausible because of the over-abundance (on either the cosmic or solar scales) of iron relative to other elements that might be expected to produce ions in the proper M/Q range at the ionization temperatures inferred from the oxygen and silicon observations. The expected envelope of the iron ionization state is indicated. The relative heights of the observed peaks are again consistent with an ionization temperature between 1.5 and $1.75 \times 10^6\,^\circ K$. The total flux of iron in these five spectral peaks gives an elemental abundance relative to ^{16}O of 0.17. This value is consistent with both the coronal and photospheric (the latter as recently revised) iron abundances [4.44, 4.45].

The spectra and discussion presented above summarize our present knowledge of the ionic composition of the interplanetary plasma. The limitations of the observations should be obvious. Identification of ionic

species based on energy per charge alone must always remain somewhat ambiguous and dependent on our prejudices (well-founded as
they may be) regarding the probable elemental composition of solar
material. Observation of ions other than $^1H^+$ and $^4He^{++}$ cannot be
performed routinely, and the measurements described above may not
represent a typical sample of the solar wind chemical composition.
Nonetheless, there is both beauty and consistency in these measurements. Even those with little interest in the details of satellite instrumentation should appreciate the *tour-de-force* of clearly resolving the five
closely spaced spectral peaks (attributed to iron) in Fig. 4.9. A consistent
implication of the observations is an ionization temperature somewhat
above $1.5 \times 10^6\,°K$. This leads quite directly to the interpretation of the
interplanetary ionization state as a physical property little changed in
transit from the sun to 1 AU (IV.8).

To these observations can be added two results concerning the interplanetary presence of ^3He and ^{20}Ne deduced from the foil trapping
technique already described in IV.2 [4.17]. The Apollo 11 and 12 foil
exposures led to average ^3He to ^4He flux ratios of 5.4×10^{-4} and
4.1×10^{-4}. These values are in reasonable agreement with that derived
from the Vela 3 spectra of Fig. 4.8. Both exposures also implied an
average ^{20}Ne flux $1.3 \times 10^4\,cm^{-2}$, giving flux ratios of ^4He and ^{20}Ne
equal to 430 and 620. The ionization state of solar wind neon is not
determined by these observations.

IV.8 The Ionization State of Expanding Coronal Plasma

In all of our previous theoretical discussions, we have written the particle
conservation equation in a steady, spherically-symmetric coronal
expansion as

$$\frac{1}{r^2}\frac{d}{dr}(n_i u_i r^2) = 0, \qquad (4.16)$$

where n_i and u_i are the density and radial expansion speed of the ith
constituent of the plasma. The more general form of this conservation
law is

$$\frac{1}{r^2}\frac{d}{dr}(n_i u_i r^2) = \sum_i (r), \qquad (4.17)$$

where $\sum_i (r)$ is a source function for the ith constituent, with the units
$cm^{-3}\,sec^{-1}$. $\sum_i (r)$ could be set equal to zero in discussing the expansion
of a corona composed either of hydrogen or of hydrogen and helium
because both of these elements are expected to be essentially completely
ionized at electron temperatures above $\sim 2 \times 10^5\,°K$, or in the transition
zone, the corona, and nearly all of interplanetary space inside of 1 AU.

However, discussion of the expansion of a corona including elements that are not completely ionized requires use of the general conservation law (4.17) and explicit consideration of ionization and recombination processes.

Consider the ion X^i (with charge $+i$) of a hypothetical element X. The dominant ionization and recombination processes involving this ion in the corona are [4.46]:

(1) *Collisional ionization by electron impact,*

$$X^i + e \rightarrow X^{i+1} + e + e,$$

where e represents an electron. This process depletes the X^i population at the rate $C_i n_i n_e$, where n_e is the electron density. The ionization coefficient $C_i = \langle \sigma_c v \rangle$, where σ_c is the ionization cross section and v is the relative speed of the ion and electrons; C_i is essentially a function of the electron temperature only.

(2) *Radiative recombination,*

$$X^i + e \rightarrow X^{i-1} + v$$

where v represents a photon. So-called dielectronic recombination, in which an electron is captured by X^i to a doubly excited state, must be included here, and is extremely important under coronal conditions [4.47]. This process depletes the X^i population at the rate $R_i n_i n_e$. The recombination coefficient $R_i = \langle \sigma_R v \rangle$, where σ_R is the recombination cross section; R_i is again a function of electron temperature only.

The processes inverse to (1) and (2), namely three-body recombination and photoionization, are unimportant at the low densities and low fluxes of photons (with sufficient energy to ionize coronal ion species) prevailing in the corona and interplanetary space. The source term for the ion X^i is then

$$\sum_i (r) = -C_i n_i n_e - R_i n_i n_e + C_{i-1} n_{i-1} n_e + R_{i+1} n_{i+1} n_e . \qquad (4.18)$$

The ionization state of element X is determined by writing a particle conservation law of the form (4.17), using the source term (4.18), for all of the ions $i = l, l+1, \ldots p$ present (or at least important) in the corona. To this set of equations must be added a momentum equation and an energy equation for each ionic species. Needless to say, the simultaneous solution of this coupled system of equations is the most formidable theoretical problem we have yet discussed.

The basic properties of the solutions to this problem are easily understood by consideration of two limiting cases. For the simple example of two ions, X^i and X^{i+1}, the particle conservation law becomes

$$\frac{1}{r^2} \frac{d}{dr} (n_i u_i r^2) = -C_i n_i n_e + R_{i+1} n_{i+1} n_e . \qquad (4.19)$$

In the limit of a vanishing expansion speed u_i, the divergence of the particle flux is negligible, so that (4.19) reduces to

$$- C_i n_i n_e + R_{i+1} n_{i+1} n_e = 0 . \tag{4.20}$$

In other words, a local balance (as in a static atmosphere) exists between the depletion and production processes. Equation (4.20) has the simple solution

$$\frac{n_i}{n_{i+1}} = \frac{R_{i+1}}{C_i} . \tag{4.21}$$

The ionization state is independent of the electron density, depending only on the local electron temperature through the coefficients R_{i+1} and C_i. In the opposite limit of a vanishing electron density, the source term becomes negligible, so that (4.19) reduces to the familiar

$$\frac{1}{r^2} \frac{d}{dr} (n_i u_i r^2) = 0 . \tag{4.22}$$

In other words, the flux $I_i = 4\pi n_i u_i r^2$ of X^i is conserved. The physical parameter that determines approach to these limiting behaviors is the ratio $\imath = \tau_e / \tau_a$, where τ_e is the time scale characteristic of coronal expansion and τ_a is the time scale characteristic of the atomic processes (ionization and recombination). The time scale τ_e has already been introduced and employed in the classification scheme of II.2 (where it was denoted by τ_1). The time scale τ_a is simply given by

$$\tau_a = \frac{1}{C_i n_e} \quad \text{or} \quad \frac{1}{R_i n_e}$$

for ionization or recombination of X^i. If $\imath \gg 1$, the atomic processes occur a time scale short compared to the expansion, and we would expect the ionization state to adjust to local conditions (as in a static atmosphere where τ_e is infinite), the first of the limiting cases described above. If $\imath \ll 1$, the atomic processes occur on a time scale long compared to the expansion, and should have little effect on the ionization state, the second of the limiting cases. This intuitive argument regarding the role of $\imath$ can be confirmed by writing (4.19) in dimensionless form.

The densities prevailing in interplanetary space imply very long scale times for atomic processes and lead us to expect that the interplanetary plasma is in the $\imath \ll 1$ realm [4.48, 4.49, 4.50, 4.51, 4.53]. As a specific example, consider the ionization state of oxygen in the corona and solar wind. At coronal temperatures, O^{+6} and O^{+7} are expected to be the most abundant ionic species [4.52]. Fig. 4.10 shows τ_6, the scale time for ionization of O^{+6}, τ_7, the scale time for recombination of O^{+7} and τ_e, as functions of heliocentric distance in the basic one-fluid coronal expansion model of Whang and Chang (III.4). Both τ_6 and τ_7 are much less than τ_e near the base of the corona, so that $\iota \gg 1$, and the ionization state of oxygen would be expected to be determined by the local electron temperature. However, the atomic scale times increase rapidly with heliocentric distance (due to their inverse dependence on the rapidly declining density), becoming equal to the expansion scale time at $r \approx 1.2 r_\odot$ and much larger than the expansion time a short distance above this level. Thus $\iota \ll 1$ for $r \gtrsim 1.5 r_\odot$, and the ionization state of oxygen would be expected to change but little beyond this position. The interplanetary ionization state must be essentially determined by conditions within $r \approx 1.5 r_\odot$, or well within the corona.

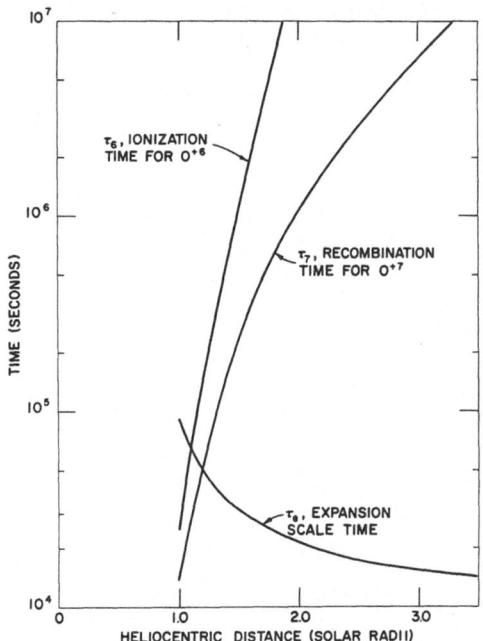

Fig. 4.10   Scale times for the atomic processes affecting the ionization state of oxygen and the scale time for expansion of the corona, as functions of heliocentric distance [4.53, 4.54]

Solutions of the particle flux conservation equations (4.17) have been obtained only under the simplifying assumptions of a specified expansion speed $u(r)$ (equal for all ions), electron density $n_e(r)$, and electron temperature $T_e(r)$. Fig. 4.11 shows (as the solid lines) the oxygen ionization state obtained by numerical integration with $u(r)$, $n_e(r)$ and $T_e(r)$ from the basic one-fluid model. The oxygen ionization state expected in a static corona with the same $T_e(r)$ is indicated by the dashed lines. As expected from the discussion of scale times, the "dynamic" solution follows the static solution near $r=r_\odot$, but beyond $r \approx 1.5 r_\odot$ displays an ionization state virtually independent of r. The relative abundance of O^{+6} and O^{+7} at large heliocentric distances is equal to that given by the static solution at $r \approx 1.2 r_\odot$, or where $\iota \approx 1$. Thus the interplanetary ionization state of oxygen is determined by conditions surprisingly deep in the corona, and is "frozen-in" during the expansion beyond this position. A similar conclusion has been reached by Kozlovsky [4.55] from integrations performed under the assumption of a constant expansion speed. In fact, the rapid attainment of the fixed ionization state ($\iota \ll 1$) in the expanding plasma implies that this result is not very model-dependent; the ionization state shown on Fig. 4.11 at large r is little different from the static state at the base of the corona.

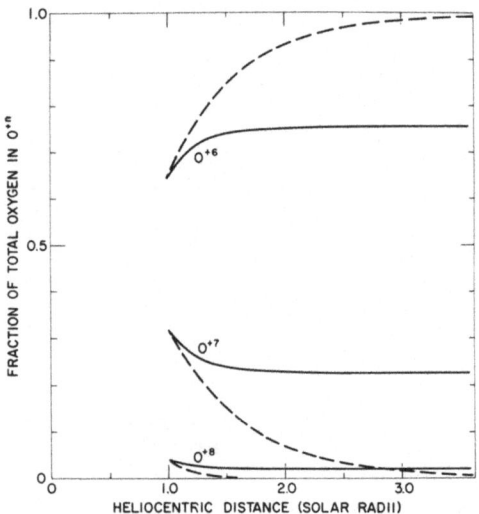

Fig. 4.11   The predicted ionization state of oxygen in an expanding (solid lines) and static (dashed lines) corona [4.53, 4.54]

The observations indicating an ionization temperature of solar wind oxygen, silicon, and iron (IV.7) of 1.5 to $1.75 \times 10^6\,^\circ K$ are easily understood in terms of the model described above. The solar wind ionization temperature should be essentially that of the coronal region from which the solar wind emanated, and solar wind observations can thus give a direct determination of that coronal temperature. This survival of coronal characteristic in the expanding plasma could prove very useful if ionization state observations were made routinely. For example, the determination of an ionization temperature representative of high speed plasma streams (II.3) might be of assistance in attempts to trace the coronal source of these solar wind features (Chapter V). More subtle analyses can be envisioned. The ionization states of different elements would be expected to "freeze" at different heliocentric positions; whereas the state of oxygen has been shown to reflect the coronal temperature near $r = 1.2 r_\odot$, that of iron can be shown to depend on conditions between $2 r_\odot$ and $3 r_\odot$. Accurate and simultaneous observations of the interplanetary ionization states of different elements might lead to a determination of the coronal temperature gradient.

IV.9 Possible Interplanetary Modification of the Solar Wind Ionization State

Although the major observed features of the interplanetary ionization state (i.e., the predominance of O^{+6} among the oxygen ions, the relative abundances of Si^{+8} and Si^{+9} and of Fe^{+8} to Fe^{+12}) can be explained by ionization and recombination processes within the expanding coronal plasma, several observed features do not fit into this concept. In particular, two small spectral peaks observed by Vela 3 have been tentatively attributed to O^{+5} and $^4He^+$ (IV.7). The observed fluxes are far larger than can be explained by the model of IV.8. Alternate identifications of these spectral features have been proposed [4.40, 4.41], so that the evidence for the presence of these ions is not completely convincing.

Nonetheless, these observations have led to the speculation that such ions could be produced in interplanetary space by interactions of solar plasma or radiation with material of interstellar origin. The first such suggestion [4.54] invoked the charge exchange process

$$H + He^{++} \rightarrow H^+ + He^+$$

to produce He^+ from solar wind He^{++} through interaction with neutral interstellar hydrogen atoms (presumed to have penetrated to within 1 AU of the sun). A similar process could produce O^{+5} from solar wind O^{+6}. Although the cross-section for this process is large, the flux of interstellar hydrogen required to produce the observed (on one occasion) He^+ flux is higher than can be reconciled with recent observations of the scattering of solar Lyman-α radiation in interplanetary space [4.56]. Holzer [4.57] has proposed the photoionization process

$$He + v \rightarrow He^+ + e$$

to produce He^+ from neutral interstellar helium atoms. Banks [4.58] has proposed a more complex process involving interstellar dust particles rather than gas. The bombardment of dust grains by the solar wind would lead to adsorption of solar wind hydrogen and helium until the outer layers of the grains became saturated. After saturation has occurred (in ~ 200 years for He), further bombardment releases neutral helium atoms. The same photoionization process suggested above could then lead to He^+ in the solar wind.

Each of these proposals employs an interaction with the interstellar medium to produce an interplanetary modification of the solar wind ionization state. It should be emphasized that these modifications are minor, involving small concentrations of ions in low charge states. The observational evidence for the existence of such ions is marginal. However, these suggestions point to a further significance of interplanetary ionization state observations—their potential use as a tool in studying the presence of and solar wind interaction with gas or dust from interstellar space. For further discussion of such interactions, the reader is directed to a recent review by Axford [4.59].

Chapter V

High-Speed Plasma Streams and Magnetic Sectors

V.1 Introduction

We have already found (III.2) that the "quiet" interplanetary conditions indicative of a structureless coronal expansion are rarely observed in the real solar wind. Mere inspection of solar wind data reveals large variations on a time scale of several days, comparable to the basic scale time of the overall coronal expansion (and thus indicative of phenomena in Class 2 of the classification scheme developed in Chapter II). Autocorrelation analysis of solar wind speed measurements confirms this conclusion. Prominent among these variations is the pattern associated with the high-speed plasma streams (II.3) first identified in the Mariner 2 data of 1962. The apparent recurrence of the streams during the period of Mariner 2 observations suggested that they were long-lived, spatial features that retained their basic identity for longer than a solar rotation. The variations observed in interplanetary space would then result from rotation of this spatial structure past the observer. The polarity of the interplanetary magnetic field has also been found to be organized into unipolar regions with a similar scale size (or time), that are related to the high-speed streams. We will now review the observed physical features of the streams and magnetic regions and discuss models of their solar origin and interplanetary evolution. The influence of high-speed plasma streams on solar wind properties inferred from long-term statistical analysis of observations as well as the role of these streams in the overall transport of mass and energy from the corona will be discussed. Finally, the relationship of the streams to solar activity and the hypothetical "M-regions" will be considered.

V.2 Observed Features of High-Speed Plasma Streams

The classical examples of high-speed plasma streams remain those observed by Mariner 2 and displayed on Fig. 1.5. The basic physical features of these streams can be summarized as follows [5.1]:

(1) The flow speed rose steeply from the pre-stream level ($\sim$325 to 400 km sec^{-1}), reaching a maximum value ($\sim$600 to 700 km sec^{-1}) in about one day. The subsequent decrease in flow speed was more gradual, with several days required for return to the approximate pre-stream level.

(2) The density rose to unusually high values (the highest observed during the entire Mariner 2 flight) near the leading edges of the streams; the high densities generally persisted for $\sim$1 day. These density peaks were generally followed by unusually low densities, persisting for several days.

(3) The proton temperature varied in a pattern similar to that of the flow speed.

These generalities have been confirmed by subsequent observations. Fig. 5.1 shows Vela 3 observations of a well-defined high-speed stream. In this particular example, solar wind properties were reasonably steady for nearly a day before arrival of the stream and the "pre-stream" conditions are thus fairly well determined. The densities observed during the slow decline in flow speed were distinctly lower than the pre-stream density. The proton temperature elevation near the leading edge of the

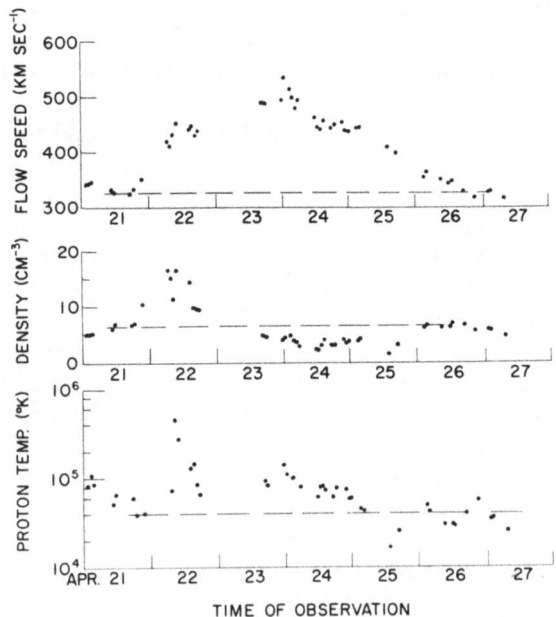

Fig. 5.1   A high-speed solar wind stream observed by Vela 3 spacecraft in April, 1966 (unpublished data, courtesy of S. J. Bame). The dashed lines represent the approximate pre-stream values of flow speed, density, and proton temperature

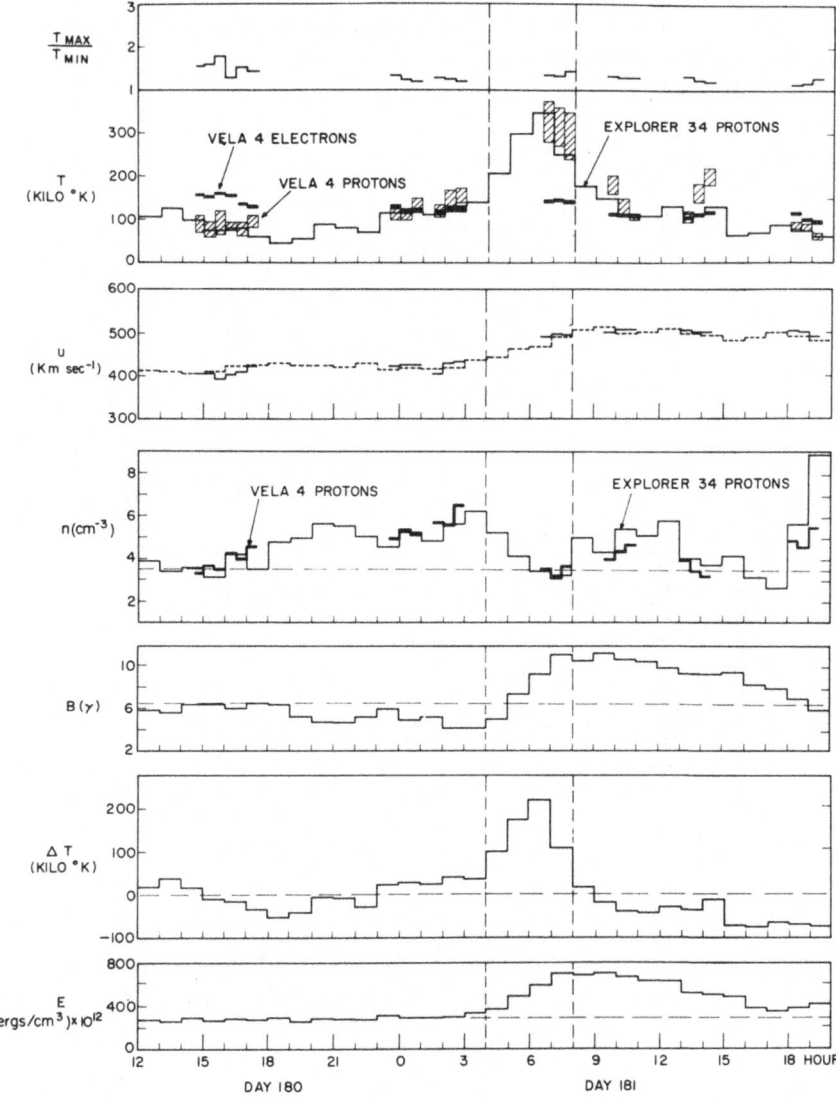

Fig. 5.2   The leading portion of a high-speed stream observed by the Vela 4 and Explorer 34 spacecraft [5.3]. T_{max}/T_{min} is the magnitude of the proton temperature anisotropy, ΔT is the difference between the observed proton temperature and the empirical $T(u)$ proposed by Burlaga and Ogilvie [5.2] (given in V.7), and E is the thermal energy density

stream has been emphasized by Burlaga and Ogilvie [5.2] and Burlaga *et al.* [5.3]. Fig. 5.2 shows observations from Vela 4 and Explorer 34 spacecraft in such a region. In this example it is particularly clear that the density, proton temperature, and flow speed reach maximum values in sequence.

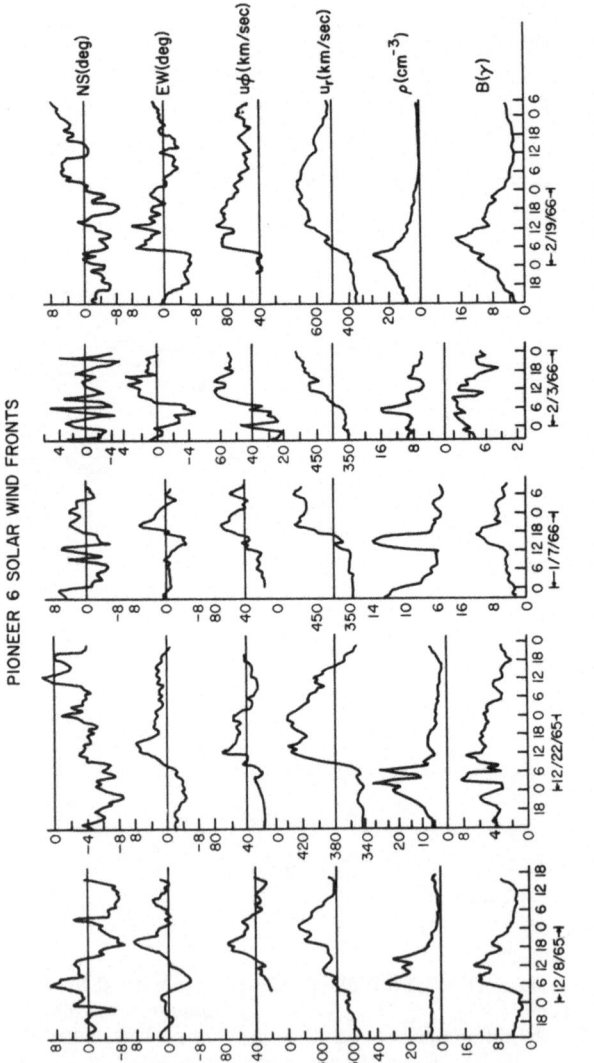

Fig. 5.3 Solar wind properties observed in several high-speed streams by the MIT plasma probe on Pioneer 6 [5.7]

Recent observations have added additional details to this basic description of high-speed streams. For example, Fig. 5.3 shows the variations of several solar wind properties, including the flow direction, in five streams observed by the Pioneer 6 spacecraft. In each of these examples, the flow direction shifted to be from east of the sun (see Fig. 3.20) in the leading edge of the stream, and then shifted to be from a more westerly direction near the maximum of the flow speed variation. This pattern has been discerned by several independent observers [5.4, 5.5]. The Vela 4 observations of electron temperature, illustrated in Fig. 5.2, revealed little or no variation within high-speed streams, even near the leading edges where the proton temperature attained its maximum value.

The tendency for high-speed streams to recur on several successive solar rotation has been less obvious in the more recent observations [5.6]. Nonetheless, Siscoe [5.7] has examined the flow velocity changes at the leading edge of streams, finding a pattern of variations consistent with the previous interpretation as spatial structures rotating with the sun. In the context of this interpretation, the time scale for passage of the *entire* high-speed stream, ~ 4 days (II.3), implies a spatial extent of $\sim 1\,\text{AU}$ ($\omega r_e \approx 400\,\text{km sec}^{-1}$). Note, however, that this scale size pertains only to the solar-longitude extent of a stream. The scale size in solar latitude could conceivably be quite different (V.5). Possible evidence for such a difference was obtained by Gosling [5.8], who compared the flow speeds observed by Pioneer space probes spaced at different solar longitudes near the orbit of the earth. The correlations among the different observations were good for closely spaced observations but became poor for observations performed at longitude differences greater than 90°. This apparent failure for speed structures to persist for more than 1/4 of a solar rotation could be explained either by time variations on an ~ 7-day time scale or by a spatial structure in solar latitude on an $\sim 1/10\,\text{AU}$ scale (equivalent to the 7° difference in solar latitude implied by a 90° difference in solar longitude in the ecliptic plane, tilted at 7° from the solar equator). The interpretation as a latitude effect is consistent with Vela 3 observations of a possible latitude dependence in 27-day averages of the solar wind density and flow speed over the same $\pm 7°$ range of solar latitude [5.9].

V.3 Observed Features of Magnetic Sectors

Proper averaging of the "magnetic polarity" observed near 1 AU reveals the existence of another large-scale pattern. The magnetic field tends to point predominantly toward or away from the sun (along the basic spiral orientation of Fig. 1.4) for intervals of several days. The further tendency for these intervals of organized polarity to recur with a period near 27 days has led to the concept of a long-lived interplanetary magnetic sector pattern that rotates with the sun.

This concept was developed by Wilcox and Ness [5.10, 5.11, 5.12] in interpreting the magnetic field measurements performed on the Imp 1 spacecraft in late 1963. Fig. 5.4 displays three-hour averages of the sign of the radial field component at the solar longitude (in a frame rotating with the sun) of observation; the plus signs denote an average field pointing away from the sun, while the minus signs denote an average field pointing toward the sun. The tendencies for large spatial regions (or temporal intervals) to have a well-defined predominant polarity and for the polarity pattern to repeat on succeeding solar rotations are clear. The boundaries between the regions of differing polarity have been schematically extended toward the sun on Fig. 5.4 in accord with the concept of spiral field lines (Fig. 1.4). Wilcox and Ness introduced the term "magnetic sectors" to describe these regions. In a stationary frame of reference the sectors would appear to rotate with the sun.

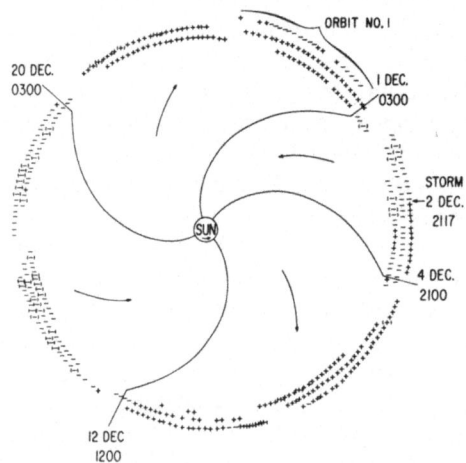

Fig. 5.4   The magnetic polarity pattern observed by the Imp 1 spacecraft [5.10]. The sign of each three-hour average value of the radial field component is shown at the solar longitude of observation

The reader will recall that the spiral magnetic field model of I.5 describes only the orientation of the field lines. The sense and intensity of the field at any point is independently determined by the field configuration on the "source surface" for the flow (the function $B(r_0, \phi, \theta)$ in equations 1.24). The interplanetary sector pattern of Fig. 5.4 then implies that the solar wind within each magnetic sector emanated from a coronal region of similarly organized polarity. These regions would be of large extent if the solar wind is hypothesized to originate in an overall (or at least widespread) expansion of the corona, or of small extent if the solar wind is hypothesized to originate in a highly non-uniform expansion from a few small coronal sources. Although both of these possibilities, referred to as the "mapping hypothesis" and the "nozzle hypothesis" are plausible, the former has found greatest acceptance.

The magnetic field of the solar photosphere (outside of the polar zones and sunspots) is observed to consist of large areas within which a single magnetic polarity predominates. These regions evolve slowly during a solar rotation. A cross-correlation analysis of the observed photospheric polarity and that observed at 1 AU by Imp 1 revealed a significant relationship with a time lag of $\sim 4\frac{1}{2}$ days [5.12]. This lag is in accord with the time required for a signal to propagate from the sun to 1 AU at the average solar wind speed of $\sim 400\,\mathrm{km\,sec^{-1}}$. Fig. 5.5 is a "synoptic chart" of the photospheric magnetic field observed between Nov. 25 and Dec. 11, 1963, with the solid contours denoting a positive polarity (outward field) and the dashed contours denoting a negative polarity (inward field). Solar longitude is replaced by the time of central meridian passage of a given feature (the observations are, in fact, made only near central meridian) in such a chart. The boundaries of the large positive magnetic sector observed by Imp 1 between Dec. 2 and Dec. 12, 1963 (in the right hand, lower portion of Fig. 5.4) are projected onto this chart by assuming a $4\frac{1}{2}$-day transit time. The large area of positive photospheric polarity in the northern solar hemisphere nearly coincides with the projected position of the interplanetary sector, suggesting that this magnetic region has been "mapped" into interplanetary space by the coronal expansion. The detailed recurrence pattern of the photospheric field also led Wilcox and Ness [5.11] to conclude that the region between 10° and 20° North solar latitude was the source of the observed interplanetary field at that time.

The earlier Mariner 2 measurements also revealed a similar magnetic sector pattern, as illustrated in Fig. 5.6. The pattern differs from that of Imp 1 in details such as the duration of the sector and the number present during each solar rotation. Subsequent observations have confirmed this general organization of the interplanetary magnetic field; Fig. 5.7 shows the polarity pattern inferred from magnetometer observations made between 1962 and 1970 [5.14]. The details of the pattern appear to have varied with changes in solar activity. The Mariner 2 and Imp 1 observations, performed during the stable, declining portion of the previous

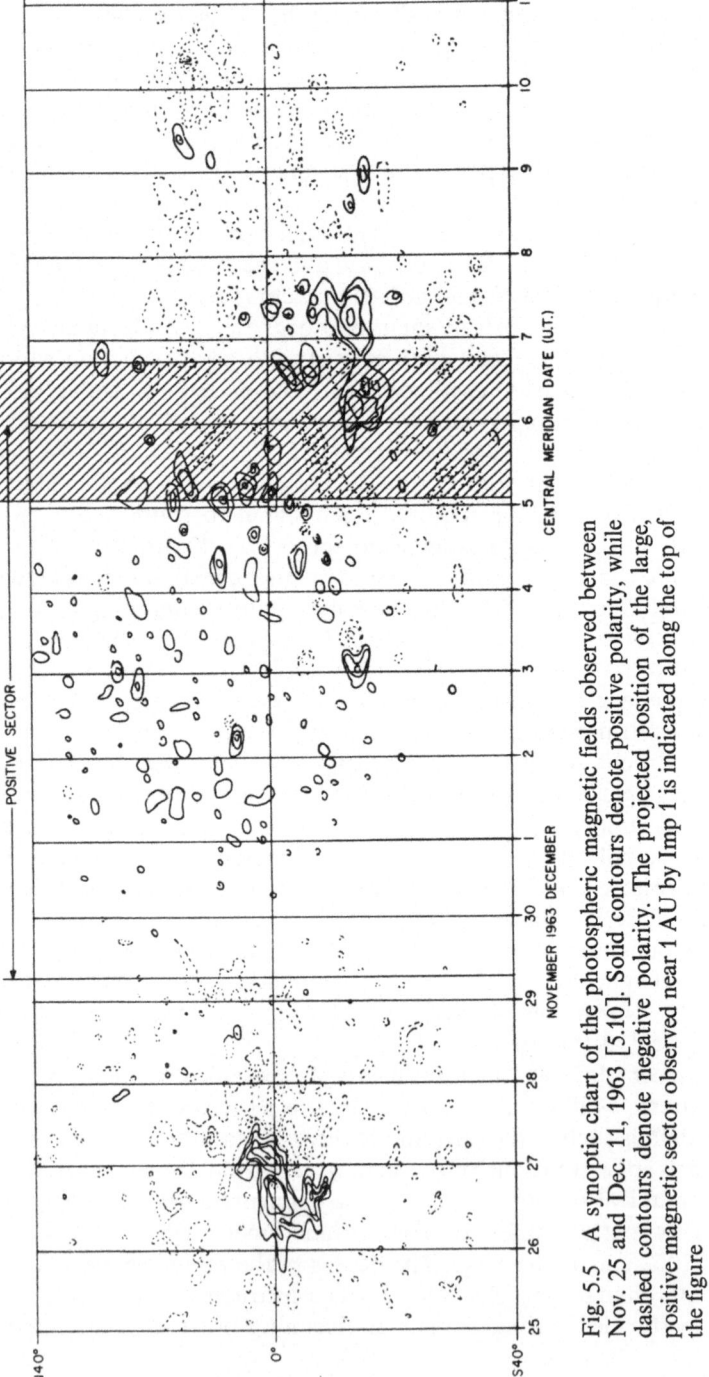

Fig. 5.5   A synoptic chart of the photospheric magnetic fields observed between Nov. 25 and Dec. 11, 1963 [5.10]. Solid contours denote positive polarity, while dashed contours denote negative polarity. The projected position of the large, positive magnetic sector observed near 1 AU by Imp 1 is indicated along the top of the figure

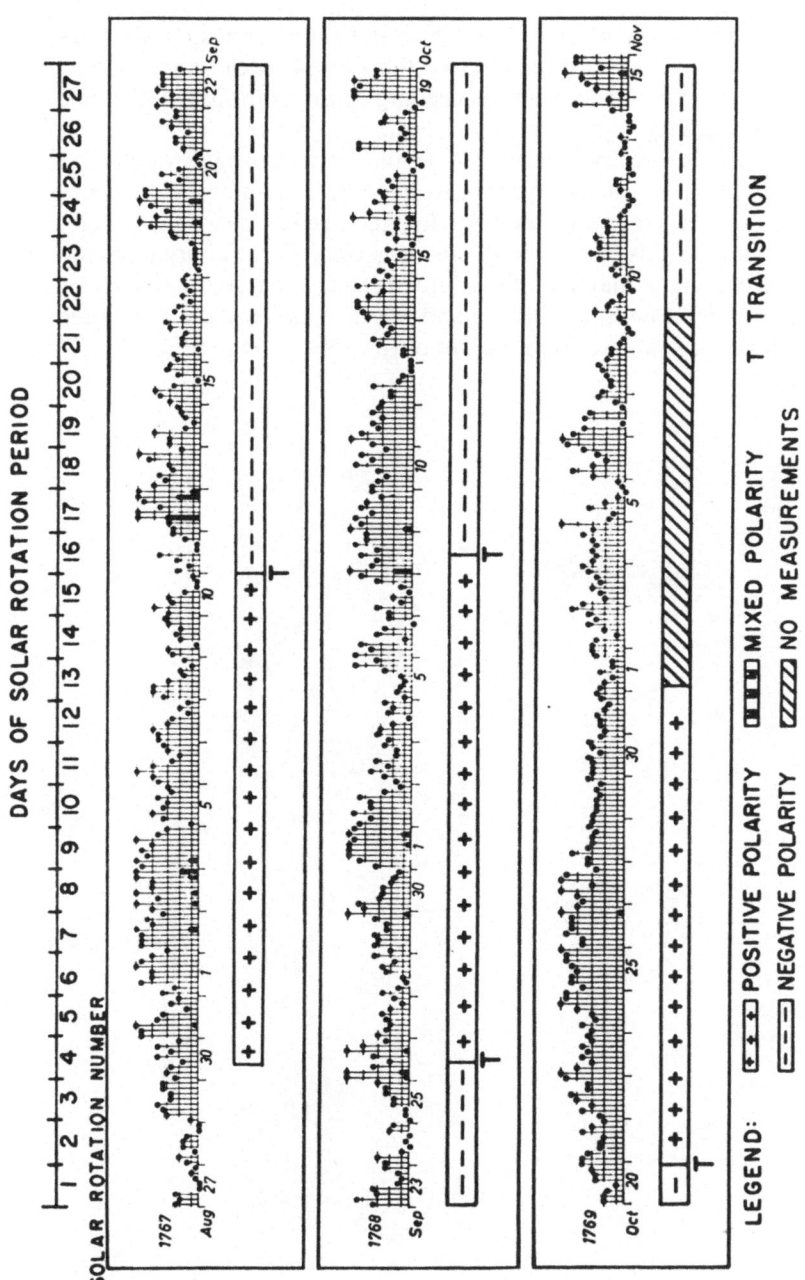

Fig. 5.6 The magnetic polarity observed by the Mariner 2 spacecraft in 1962 [5.13]. The K_p index of geomagnetic activity is also given

solar cycle, revealed a steady, recurrent polarity pattern. Observations performed during the rising portion of the present cycle (e.g., 1965 to 1967) showed greater changes on succeeding solar rotations, with simple recurrences occuring only rarely. The basic concept of rotating, nearly stationary sectors, as drawn in Fig. 5.4, would be only roughly applicable to this interval. The pattern appears to have become more stable since mid-1967, with a renewed tendency for recurrence over many solar rotations. Despite this change in details it is clear that the organization of the interplanetary magnetic field into sectors, each characterized by a well-defined predominant polarity and requiring several days to sweep past an observer, is a persistant feature of the coronal expansion.

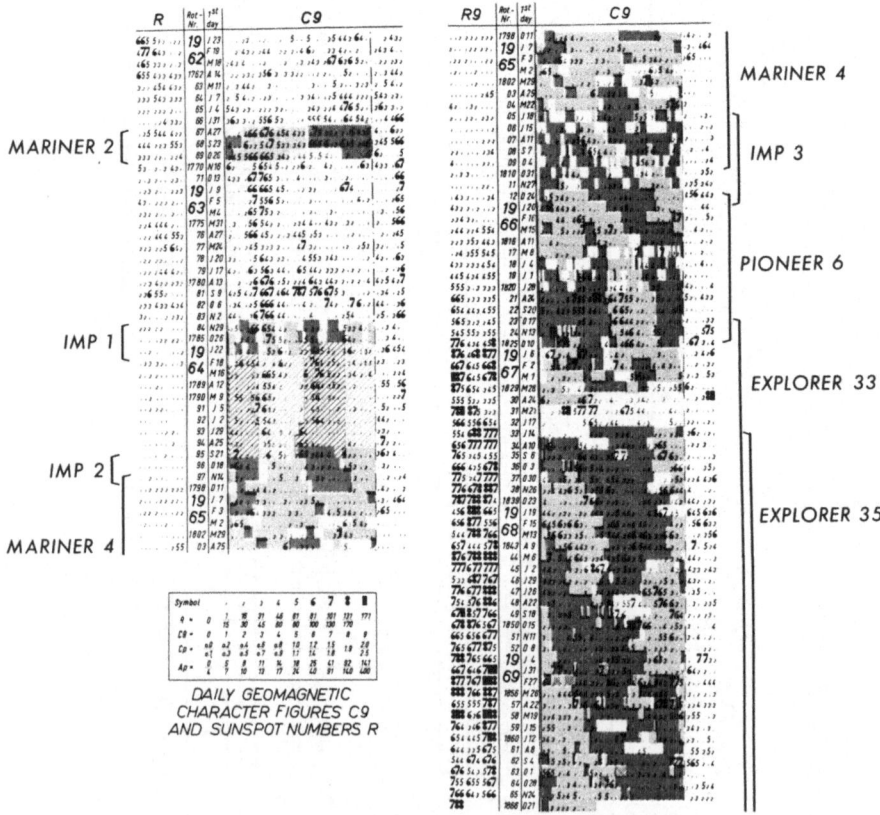

Fig. 5.7 The magnetic sector pattern observed in interplanetary space from 1962 through 1969 [5.14]. The light shading indicates positive polarity while the dark shading indicates negative polarity

V.4 The Relationship between High-Speed Plasma Streams and Magnetic Sectors

As the high-speed plasma streams (V.2) and magnetic sectors (V.3) show similar time scales and recurrence tendencies, it is not surprising to find that they are related. Fig. 5.8 displays three-hour averages of the flow speeds observed by the MIT plasma probe on Imp 1 within each magnetic sector [5.10]. The flow speed was low near the sector boundary, rising to a maximum value one to two days later, and then dropping to a low value five to seven days into the sector. This is a near repetition of our earlier description of a high-speed stream. The compression and rarefaction in density were also revealed by averaging the plasma observations within the Imp 1 sectors [5.10]. This led to the conclusion that the high-speed streams and magnetic sectors observed by Imp 1 were essentially identical.

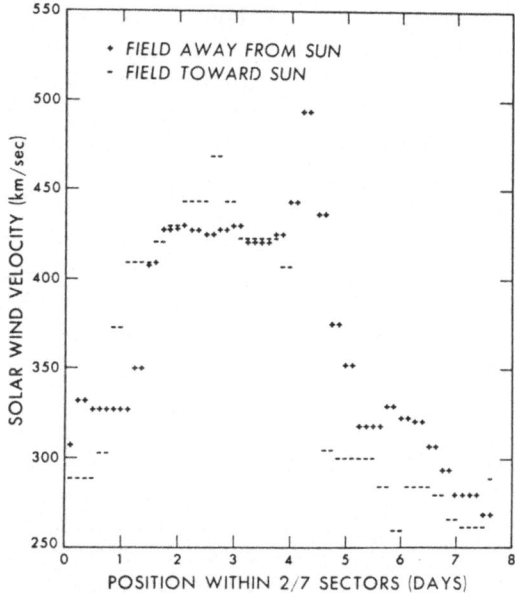

Fig. 5.8   A superposed plot of three-hour averages of the solar wind speed observed within the Imp 1 magnetic sectors [5.10]

The Mariner 2 observations revealed a slightly different relationship between the plasma and magnetic field. Inspection of Figs. 1.5 and 5.6 reveals that two high-speed streams occurred within the large negative polarity regions observed during the last ~ 12 days of each solar rotation. Thus the identity of high-speed streams and magnetic sectors is not maintained. However, the more limited conclusion that each important high-speed stream has a predominant magnetic polarity appears to remain valid.

Subsequent observations have generally confirmed this magnetic characteristic of well-defined high-speed streams. Ness *et al.* [5.5] have found that the organization of Vela 3 plasma observations within magnetic sectors observed by Imp 3 in 1965 and 1966 confirms the general pattern of flow speed and density variations discovered by Imp 1. The variations in proton temperature and flow direction found in high-speed streams (V.2) also appeared within the sectors in this analysis. These patterns were discernible for most well-defined individual sectors, although not present in a few examples. In view of the rapid evolution of the magnetic sector pattern during this time interval (Fig. 5.7), the latter result may be less surprising than the general validity of the stream-sector relationship.

The tendency for high-speed streams to have a predominant magnetic polarity implies an origin in a coronal region with a predominant polarity. The expansion of plasma from any such region would transport the field lines into interplanetary space, retaining the identity of plasma and field lines because of the "frozen-in" nature of the field. In interplanetary space, the kinetic energy density of the plasma is ~ 100 times the energy density of the magnetic field, so that the latter is clearly relegated to a passive role. However, deep in the corona, the magnetic energy density can exceed both the thermal and kinetic energy densities, and the magnetic field could channel the coronal expansion and strongly influence the properties of the solar wind. The existence of a predominant magnetic polarity within high-speed plasma streams appears to be evidence that the coronal magnetic structure has governed the formation of these solar wind features.

V.5 Theoretical Models of Interplanetary Plasma Streams

The motions of solar wind plasma streams with differing expansion speeds are strongly influenced by the effects of solar rotation. Fig. 5.9 illustrates a uniform coronal expansion at radial speed u_1 in frames of reference that are (a) stationary, and (b) rotating with the sun. In the latter frame the flow streamlines are drawn into spirals, leading to the basic configuration of the frozen interplanetary magnetic field already derived in I.5. At any given radius r in the solar equatorial plane, the angle between a streamline and the radial direction is Φ_1, where

$$\cos^2 \Phi_1 = \frac{1}{1 + \left(\dfrac{\omega r}{u_1}\right)^2}$$

(as in III.10 ω is the angular velocity of solar rotation). Fig. 5.10 illustrates, in the rotating frame of reference, the expansion of a single localized plasma stream with $u_2(r) > u_1$ into the uniform background of Fig. 5.9.

(a) STATIONARY FRAME OF REFERENCE

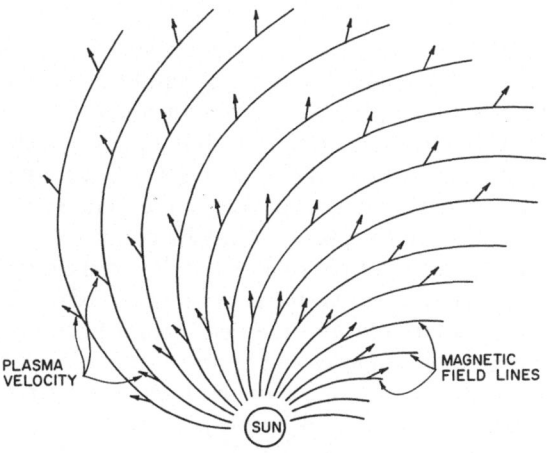

(b) ROTATING FRAME OF REFERENCE

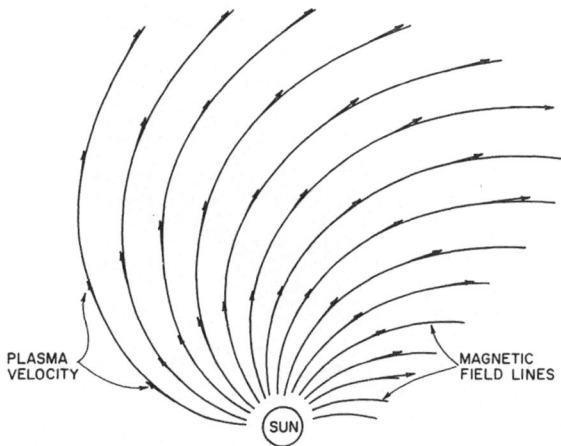

Fig. 5.9 The velocity and magnetic field configurations for a uniform coronal expansion near the solar equatorial plane. The arrows indicate the plasma flow and the light lines the magnetic field [5.24]

If undisturbed by any interaction, the streamlines for this plasma stream would be spirals making the local angle with the radial (again near the solar equatorial plane), Φ_2, where

$$\cos^2 \Phi_2 = \frac{1}{1 + \left(\dfrac{\omega r}{u_2}\right)^2}$$

As $\cos^2 \Phi_2 > \cos^2 \Phi_1, \Phi_2 < \Phi_1$, and the streamlines for the ambient flow must intersect the less tightly wound streamlines in the high-speed plasma. In other words, the high-speed stream, distorted into a spiral by solar rotation, will inevitably overtake and "collide" with any slower-moving ambient solar wind. The high electrical conductivity of the magnetized interplanetary plasma prevents interpenetration of different streams, so that the ambient plasma must be compressed and deflected (in the direction of solar rotation) to flow parallel to the interface with the fast stream. This interaction would also be expected to deflect (in the direction opposite solar rotation) the fast stream from its unperturbed motion, producing some compression of the plasma near its leading edge. The compressed ambient plasma and the high-speed stream would be separated by an interface crossed by no plasma flow or magnetic field lines—in other words, by a tangential discontinuity (II.3). If the relative velocity of the ambient and high-speed plasma is sufficiently large, a

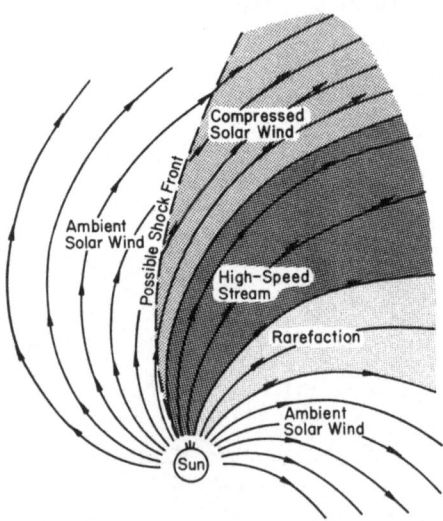

Fig. 5.10   The interaction (in equatorial cross section) of a steady, localized stream of high-speed plasma with the ambient solar wind of Fig. 5.9, as viewed in a frame of reference rotating with the sun

shock front would be expected at the leading edge of the compressed shell. Similar arguments suggest that a rarefaction must form behind the high speed stream. If conditions at the sun are assumed to remain constant, the steady (but not spherically-symmetric) flow pattern sketched in Fig. 5.10 must develop. Fig. 5.11 shows this pattern in a stationary frame of reference, in which the high-speed stream becomes a wave that propagates counterclockwise about the sun.

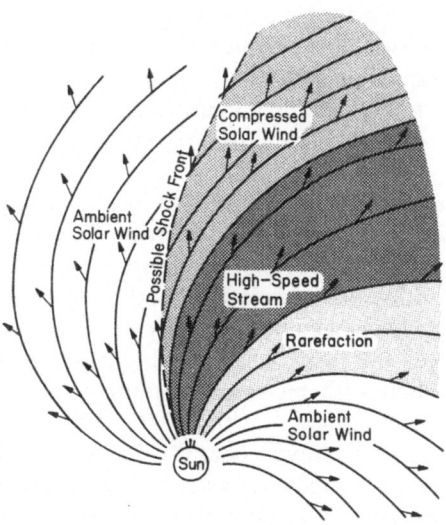

Fig. 5.11   The interaction of Fig. 5.10, viewed in a stationary frame of reference

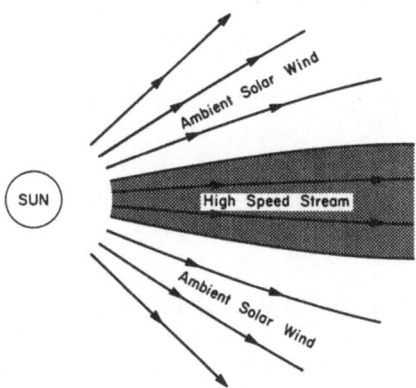

Fig. 5.12   A meridional cross section of the interaction of a localized high-speed stream with the ambient solar wind

The relative motion of the plasma streams sketched in Figs. 5.10 and 5.11 is entirely an effect of solar rotation. No such effect is present in the direction normal to the solar equator. Fig. 5.12 illustrates a plausible interaction between a high-speed stream and an ambient solar wind in a meridional plane. The high-speed and ambient plasmas may be deflected in response to any pressure differences; Fig. 5.12 has been drawn for the *possible* case in which the pressure decreases more rapidly with heliocentric distance in the stream than in the ambient. If there are no instabilities or irregularities in the flow, the plasma will move smoothly along the interface between the two flow regimes. The inevitable tendency toward relative motion or "collision" produced by solar rotation and sketched in Figs. 5.10 and 5.11 is not present in a meridional plane. Thus a high-speed stream could have quite different dimensions in solar longitude and latitude.

In any quantitative treatment of the fluid interaction described above, the spatial nonuniformity of the flow introduces new independent variables into the fluid conservation equations, adding greatly to the mathematical difficulties inherent in the system. Because of this added complexity, solutions have been derived only under the simplest physical assumptions. Whereas the steady, spherically-symmetric coronal expansion has been discussed with inclusion of numerous forces and energy sources (Chapter III), treatments of the particular steady, non-uniform flow under discussion here have generally included only the pressure gradient and gravitational forces and assumed an adiabatic or polytropic flow. Under these assumptions the mass and momentum conservation equations become (in the stationary frame of reference)

$$\frac{\partial \rho}{\partial t} + \nabla \cdot \rho \boldsymbol{u} = 0 \tag{5.1}$$

and

$$\rho \frac{\partial \boldsymbol{u}}{\partial t} + \rho(\boldsymbol{u} \cdot \nabla)\boldsymbol{u} = -\nabla P - \rho \frac{GM_\odot}{r^2}. \tag{5.2}$$

The polytropic law is simply

$$\frac{\partial}{\partial t}\left(\frac{P}{\rho^\alpha}\right) + (\boldsymbol{u} \cdot \nabla)\left(\frac{P}{\rho^\alpha}\right) = 0 \tag{5.3}$$

(or P/ρ^α is constant for any fluid element). We will be looking for solutions that are steady in the rotating frame of reference, or are functions of the three independent variables r, $\eta = \phi - \omega t$ (solar longitude in a rotating frame), and θ, the solar latitude.

The pioneering treatment of this problem was carried out by Carovillano and Siscoe [5.15], who considered *small* perturbations of a steady, spherically symmetric expansion. If attention is restricted to motions in the solar equatorial plane, the *linearized* mass and momentum equations in the stationary frame become:

$$\left(\frac{\partial}{\partial r} - \frac{\omega}{u_0}\frac{\partial}{\partial \eta}\right)\left(\frac{\rho_1}{\rho_0}\right) + \frac{\partial}{\partial r}\left(\frac{v_{1r}}{u_0}\right) + \frac{1}{ru_0}\frac{\partial v_{1\phi}}{\partial \eta} = 0; \qquad (5.4)$$

$$\left(\frac{\partial}{\partial r} - \frac{\omega}{u_0}\frac{\partial}{\partial \eta}\right)(u_0 v_{1r}) + \frac{\partial}{\partial r}\left(c_s^2 \frac{\rho_1}{\rho_0}\right) = 0; \qquad (5.5)$$

$$\left(\frac{\partial}{\partial r} - \frac{\omega}{u_0}\frac{\partial}{\partial \eta}\right)(r v_{1\phi}) + \frac{c_s^2}{u_0}\frac{\partial}{\partial \eta}\left(\frac{\rho_1}{\rho_0}\right) = 0. \qquad (5.6)$$

where ρ_0 and u_0 are the mass density and radial expansion speed in the steady, spherically-symmetric ambient flow, ρ_1 and v_1 are the non-uniform perturbations of this flow, and c_s^2 is the sound speed in the ambient medium. The linearization of (5.4) to (5.6) is based on the assumptions that $\rho_1 \ll \rho_0$ and $|v_1| \ll u_0$. Solutions of equations (5.4) to (5.6) and the polytropic law (5.3) can be derived by standard separation of variable techniques. Boundary conditions involve specification of periodic flow parameters at the inner boundary of the domain of interest, $r = r_0$, and the requirement of the same periodicity at large r. Fig. 5.13 shows the variations predicted at 1 AU for a small, sinusoidal perturbation of the radial expansion speed at $r = 0.1$ AU (i.e., r_0 is chosen to be beyond

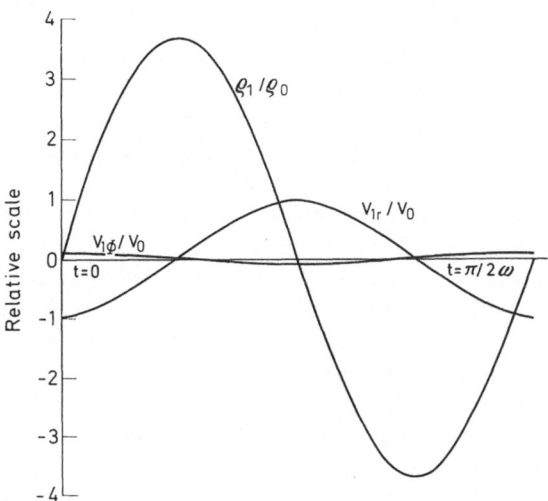

Fig. 5.13   The linear perturbations of a steady, spherically-symmetric ambient solar wind produced at 1 AU by a localized, high-speed stream introduced at 0.1 AU [5.15]

the critical point so that the flow is supersonic everywhere in the domain of interest) in an isothermal fluid. The abscissa is the time in a stationary frame of reference (Fig. 5.11) or solar longitude from 0 to 90° (i.e., there are four such waves around the solar equator) within the stationary structure in the rotating frame of reference (Fig. 5.10). A density compression and rarefaction have evolved on the leading and trailing portions of the sinusoidal velocity variation present at 1 AU. A small azimuthal velocity perturbation $v_{1\phi}$ has developed. The low speed solar wind, originally flowing in a purely radial direction, has been deflected to produce a slight "corotation," while the high-speed solar wind has been deflected in the opposite sense. The general nature of these perturbations agrees with those expected on the basis of the physical arguments presented above (and were anticipated by several authors [5.1, 5.16, 5.17, 5.18] on the basis of similar arguments). They also agree qualitatively with the observed variations in a high-speed plasma stream.

The analysis of Carovillano and Siscoe has been generalized to consider motions in latitude (but not longitude) [5.19], motions in both longitude and latitude produced by velocity perturbations at $r = r_0$ [5.20], and the general motions produced by arbitrary perturbations at $r = r_0$ [5.21]. The latter study led to a significant new physical implication. The high-speed stream structure produced at 1 AU was found to be most strongly influenced by temperature perturbations at $r = r_0$. The phases of the predicted density, flow speed, and temperature variations produced at 1 AU by an initial perturbation of the temperature are in good agreement with observations.

The linearization implicit in the models described above limits their detailed applicability to the high-speed plasma streams actually observed in the solar wind near 1 AU. It is clear that many of the streams discernible in Fig. 1.5, as well as the specific example in Fig. 5.1, involve variations in all fluid parameters that are as large or larger than the ambient values of these parameters. The observations strongly suggest nonlinear phenomena. The nonlinear evolution of the perturbations shown in Fig. 5.13 can be qualitatively predicted. The symmetric radial velocity perturbation should steepen, producing a steep rise and a more gradual decline. The density compression should become more extreme but of shorter duration, confined to the steepened rising portion of the radial speed variation. Similarly, the rarefaction should become less pronounced but of longer duration. The azimuthal velocity variations should also be distorted by the steepening of the radial velocity profile. All of these predicted changes would improve the agreement of the model with observations.

The nonlinear evolution of a localized high-speed stream has been analyzed only recently. Goldstein [5.22] has integrated the nonlinear equations (5.1) to (5.3) for motions in the solar ecliptic plane. Fluid parameters are again specified as periodic functions of solar longitude at $r = r_0 = 10\,r_\odot$ (in the supersonic flow regime) and required to be periodic at large r. Fig. 5.14 shows the evolution of the high-speed stream produced

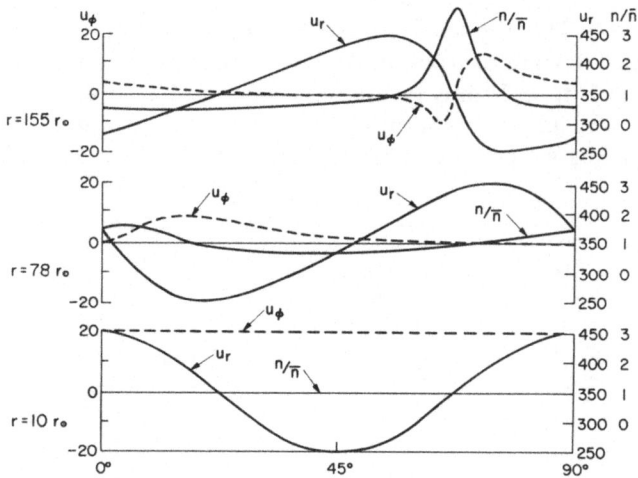

Fig. 5.14   The nonlinear evolution of a steady, localized, high-speed solar wind stream [5.22]

by a pure velocity perturbation (similar to that assumed by Caravillano and Siscoe) at r_0 (the bottom frame). The abscissa is solar longitude in a rotating frame of reference; the equivalent time scale for a stationary observer would run from right to left (the opposite convention from Fig. 5.13). The change in heliocentric longitude of the high-speed stream location at different heliocentric distances merely reflects the basic spiral configuration shown in Fig. 5.10. The changes in the shapes of the density and velocity profiles are consistent with the qualitative expectations outlined above. At $r = 155\,r_\odot \approx 3/4\,\mathrm{AU}$, the steepened velocity profile and the short, extreme compression followed by a long, shallow rarefaction are similar to the profiles observed in the solar wind. Different initial perturbations produce disturbances that differ from that of Fig. 5.14 mainly in details. Matsuda and Sakurai [5.23] have performed a similar analysis that includes the magnetic force but neglects the gravitational force, integrating the resulting equations at heliocentric distances large enough so that the sonic and Alfvén Mach number are both much larger than one. The predicted nonlinear evolution of high-speed streams is essentially the same as that displayed in Fig. 5.14.

An alternate nonlinear approach was introduced in discussing the observed elevation of proton temperatures in high-speed streams [5.3]. This treatment attempts to approximate the high-speed stream by computing the transient adiabatic motion of a spherically symmetric disturbance, a technique originally developed to study the propagation of flare-produced interplanetary shock waves (and to be further discussed in Chapter VI). The application of such a model to rotating streams is valid for small deviations from radial flow, as the conservation equations differ from (5.1) to (5.3) only (again limiting attention to motions in the solar equatorial plane) by terms proportional to u_ϕ. Fig. 5.15 shows fluid parameters as a function of time near the leading edge of a dis-

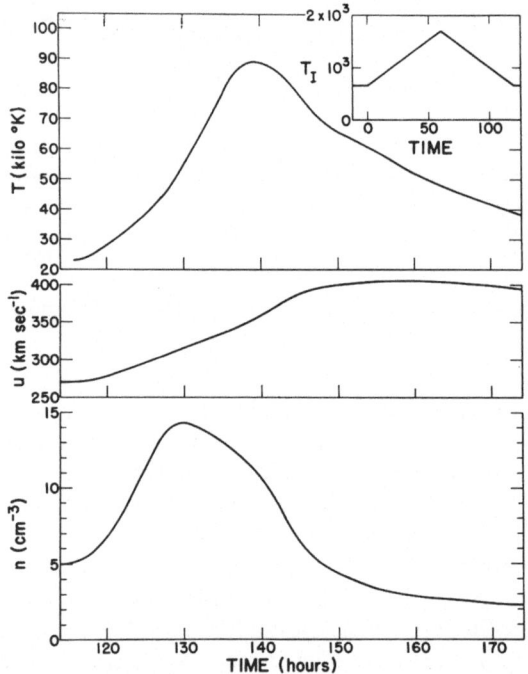

Fig. 5.15   A nonlinear, transient, spherically symmetric, solar wind disturbance produced at 1 AU by a transient temperature perturbation at 0.1 AU (inset at upper right) [5.3]

turbance produced at 1 AU by a perturbation of the temperature at $r = r_0 = 1/10$ AU (shown in the insert at the upper right). These variations in fluid parameters are again similar to those observed in the solar wind. In particular, the agreement of the predicted and observed proton temperature profiles led Burlaga et al. [5.3] to conclude that the heating of protons in such a region can be accounted for by simple adiabatic compression (actually the conversion of kinetic energy of the high-speed stream into thermal energy), with no necessity for wave dissipation or instabilities.

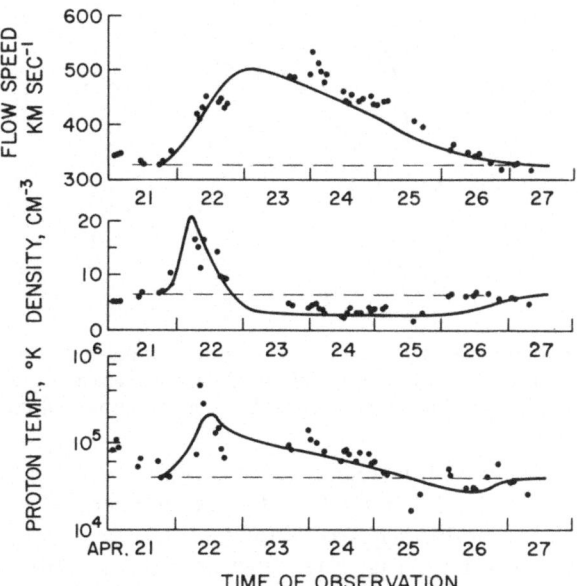

Fig. 5.16 A nonlinear, transient, spherically-symmetric solar wind disturbance at 1 AU (as in Fig. 5.15). The Vela 3 observations of Fig. 5.1 are shown for comparison

Fig. 5.16 shows an entire solar wind disturbance predicted by this transient model, directly compared with the Vela 3 observations already displayed in Fig. 5.1. The amplitude and duration of the temperature perturbation at $r_0 = 0.13\,\text{AU}$ were adjusted to produce a reasonable agreement between the predicted and observed profiles. It may be surprising that the predicted variations in density, flow speed, and temperature all agree quite well with the observations. In particular, the observed density variations are reasonably matched without assuming any density (or flux) perturbations at the source of the high-speed stream. Fig. 5.17 shows the configuration of the rotating high-speed stream equivalent to the disturbance of Fig. 5.16 (obtained by equating ϕ and ωt). Contours of the flow speed at $335\,\text{km}\,\text{sec}^{-1}$ and $475\,\text{km}\,\text{sec}^{-1}$ essentially define the extent and peak of the velocity variation. The advance of the peak speed from the center of the stream near the sun toward the front of the stream near 1 AU is again the nonlinear steepening effect, producing in

turn the density compression at the leading edge and the broad rarefaction in the trailing portion of the disturbance.

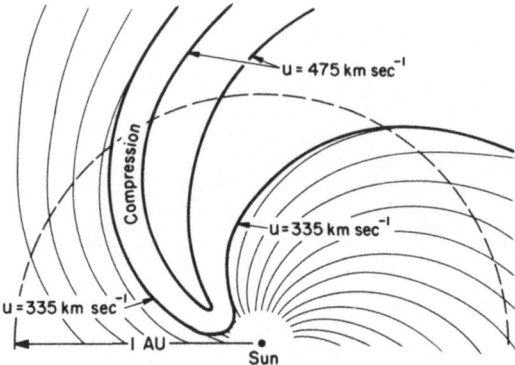

Fig. 5.17   The spatial configuration of the corotating analog of the disturbance
shown in Fig. 5.16, as given by contours of the 335 and 475 km sec^{-1} velocity levels.
The light spirals represent the magnetic field lines in the ambient flow

An expected result of the evolution of a high-speed stream would be
the formation of a shock front. Of the three nonlinear models described
above, only the latter, the approximation by a transient calculation, has
included this possibility. Fig. 5.18 illustrates the location of both a
forward and reverse shock in the model high-speed stream of Figs. 5.16
and 5.17. These shocks form at the leading and trailing edges of the
density compression in the vicinity of 1 AU. In fact, the reverse shock
appears at a smaller heliocentric distance (although at nearly the same
longitude) than does the forward shock. The indicated position of
shock formation is consistent with the observation that shock fronts are
not observed in most high-speed streams (including those of Figs. 5.1 and

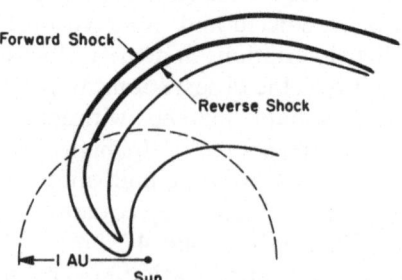

Fig. 5.18   The locations (heavy lines) of forward and reverse shock waves in the
stream of Fig. 5.17

5.16) at 1 AU [5.24, 5.25]. In fact, the approximation that u_ϕ is negligible, implicit in this model, is poorest near the shock front; inclusion of u_ϕ in the computation should reduce the density compression and move the site of shock formation still farther from the sun.

All of the theoretical models of high-speed streams described above have neglected heat conduction and the "two-fluid" nature (III.5) of the interplanetary plasma. The thermal relaxation time (that required for heat conduction to smooth out any temperature perturbation) in a conducting medium is easily shown to be [5.16, 5.26] $\tau_r \approx n k \lambda_c^2 / \kappa$, where λ_c is the scale length of the thermal gradient and κ is the thermal conductivity. If $\lambda_c \approx 1\,\mathrm{AU}$ (the scale size of a stream), then $\tau_r \approx 4 \times 10^4\,\mathrm{sec}$ or 10 h. As it requires several days for a high-speed stream to travel 1 AU, it is clear that the diffusion of energy by heat conduction is extremely efficient, and would be an important energy loss mechanism in a high-speed stream. Thus the electron temperature would not be expected to be greatly modified by the compressional heating that occurs in a high-speed stream. In contrast, the much lower thermal conductivity of the protons and their weak coupling to the electrons allows the proton temperature to be raised significantly. The observation of normal electron temperatures in the regions of highest proton temperatures associated with high-speed streams (V.2) suggests that some such rapid diffusion of electron thermal energy, quite possibly by heat conduction, does occur in the solar wind [5.26].

The quantitative models described above add substance to our earlier interpretation of a long-lived, high-speed solar wind stream as a wave when viewed in a stationary frame of reference. The spatial configuration associated with the stream propagates counterclockwise around the sun, as shown in Fig. 5.11. The plasma motion is essentially radial from the sun, so that the structure is analogous to a transverse wave. We emphasize once again that high-speed streams are nonlinear phenomena, so that the profiles of the perturbations from ambient properties and the relative phase of different perturbations are functions of heliocentric distance. The nonlinear steepening of the radial velocity profile will assume great importance in our forthcoming discussion of the solar origin of high-speed streams.

V.6 Theoretical Models of the Nonuniform, Hydromagnetic Coronal Expansion

The theoretical models described in V.5 treated only the interplanetary evolution of high-speed streams given as an initial perturbation on some source surface near the sun. These models tell us little about the origin of the streams in the nonuniform expansion produced by realistic, physical, coronal inhomogeneities, or about the physical connection between plasma streams and magnetic sectors. Models relevant to these problems must explicitly treat the interaction of plasma and magnetic fields in the corona, where magnetic forces are strong enough to have an important influence on the plasma motion. We will now describe two formulations of such models. The first treats the interaction in accurate detail, but for the sake of tractability has been limited to simple magnetic configurations. The second formulation involves a rough approximation to the inter- action, but can be applied to more complex and realistic field con- figurations.

The accurate description of a steady coronal expansion in the presence of a strong magnetic field (the first of these formulations) requires inte- gration of the mass, momentum, and energy conservation equations, along with Ampere's law relating the field to the current density j. A boundary condition involving the magnetic field is required in addition to those involving the plasma (discussed in III.4 and III.5). This additional boundary condition is usually provided by specification of the radial field component on the reference surface $r=r_0$, or at the base of the model corona. The solution of this complex boundary value problem has been approached in a series of approximations for the expansion in a dipole field and in the vicinity of coronal streamers [5.27, 5.28, 5.29, 5.30, 5.31]. We will describe in detail only the most recent and sophisticated of these analyses.

Pneuman and Kopp [5.31] have considered the mass and momentum conservation equations (5.1) and (5.2) for an isothermal corona with a dipole magnetic field at $r=r_0=r_\odot$. The neglect of solar rotation (a valid assumption near the sun) insures that the flow streamlines coincide with the magnetic field lines. The mass conservation equation can be written along a flux tube centered on any streamline as

$$\frac{d}{ds}(\rho u A(s))=0,\qquad (5.7)$$

where s is the arc length along the streamline and $A(s)$ is the (variable) cross section of the flow tube. The momentum conservation equation along a streamline is

$$\rho u \frac{du}{ds} = -\frac{dP}{ds} - \frac{GM_\odot\rho}{r^2}\cos\xi,\qquad (5.8)$$

where ξ is the angle between the streamline and the radial direction. The magnetic force $j \times B$ has no component along a streamline (parallel to B) and thus does not appear in (5.8). The momentum equation transverse to a streamline is obtained by taking the vector cross product of B with the analogue of (5.2) to obtain

$$\left[\rho(u\cdot\nabla)u+\nabla P+\frac{GM_\odot\rho}{r^2}\hat{i}_r\right]\times B=\frac{i}{c}(j\times B)\times B=\frac{i}{c}B^2 j, \quad (5.9)$$

where $\hat{i}_r$ is a radial unit vector and $B=|B|$. Equation (5.9) can be manipulated into the form

$$j=-\frac{c\hat{i}_z}{B}\left[\frac{\partial P}{\partial z}+\frac{GM_\odot\rho}{r^2}\sin\xi+\frac{\rho u^2}{R}\right], \quad (5.10)$$

where $\hat{i}_z$ is a unit vector normal to a streamline and in the plane of the assumed magnetic dipole axis, $\partial P/\partial z$ is the pressure gradient along $\hat{i}_z$, and R is the local radius of curvature of the streamline; (5.10) simply specifies the current density required to balance the transverse pressure gradient, gravitational, and centrifugal forces. Equations (5.7) and (5.8) can be integrated numerically for the plasma motion in any specified field configuration. Equation (5.10) can then be used to compute the current distribution required to maintain flow along these streamlines. Ampere's law (neglecting the displacement current)

$$\frac{c}{4\pi}\nabla\times B=j,$$

then gives the field configuration consistent with the derived current distribution. Pneuman and Kopp solved the entire system by assuming a field configuration, solving the equations as outlined above, ending the cycle with a new, derived field configuration. The latter is used to initiate a new cycle, and the process repeated until there is little change in the field assumed and derived in a single such iteration.

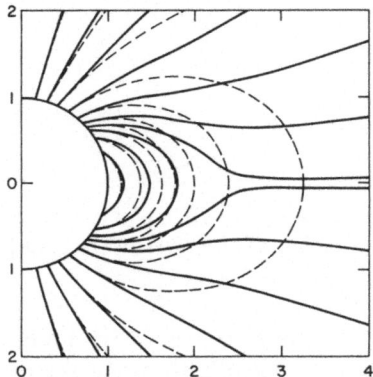

Fig. 5.19 The magnetic field line configuration for a magnetic dipole (dashed lines) and an expanding isothermal corona with a dipole field at $r=r_\odot$ [5.31]

Fig. 5.19 shows as solid lines the final solution for the field line-streamline configuration in a plane containing the dipole axis. Pure dipole field lines are indicated by the dashed lines for comparison. The pressure of the coronal plasma dominates at higher dipolar latitudes, forcing the opening of field lines and permitting expansion of the plasma. Some closed field lines remain near the dipolar equator; the plasma must be in static equilibrium in this region. Even these closed field lines are, however, somewhat distended from the pure dipole form by the plasma pressure. The transition from open to closed lines in the dipolar equator occurs at the position where, on the streamline just outside of the closed region, the flow speed equals the Alfvén speed. A current must flow along the boundary of the closed region and along the equator in the open region to maintain pressure balance.

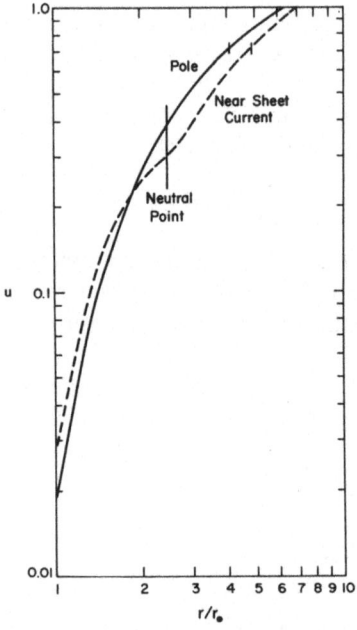

Fig. 5.20   The coronal expansion speed $u(r)$, in units of $(2kT/m)^{\frac{1}{2}}$, near the dipolar axis and near the dipolar equator given by the model of Pneuman and Kopp [5.31]

Fig. 5.20 shows the expansion speed (in units of $[2kT/m]^{\frac{1}{2}}$) as a function of heliocentric distance both along the dipole axis and on an open field line near the current sheet at the dipolar equator. For $r \gtrsim 2r_{\odot}$, the expansion speed is higher near the dipolar axis. In the context of this particular model, the coronal magnetic field has two important effects; it both "shuts off" the expansion of some coronal regions (on the closed field lines) and produces a nonuniform expansion at large heliocentric distances (well beyond the outermost closed field line). This latter effect in an isothermal model must result from the channeling of the expansion by the magnetic field.

The approximate description of a steady coronal expansion in the presence of a strong magnetic field (the second of the formulations to this problem) introduces an extreme simplification. In a spherical shell $r_0 < r < r_s$, the current density is assumed to be zero. Thus $\nabla \times B = 0$ and B can be derived from a scalar magnetic potential. For $r > r_s$, the magnetic field is assumed to be drawn into the familiar spiral pattern by the expanding plasma. In essence, this model replaces the varying influence of the plasma on the magnetic configuration, manifested by electrical currents, by assuming no influence for $r < r_s$ and dominant influence for $r > r_s$. Solution of the potential problem in the shell $r_0 < r < r_s$ requires proper specification of boundary conditions. Any distribution of the radial field component B_r can be assumed at $r = r_0$; in fact, actual photospheric magnetic observations can be used to specify a

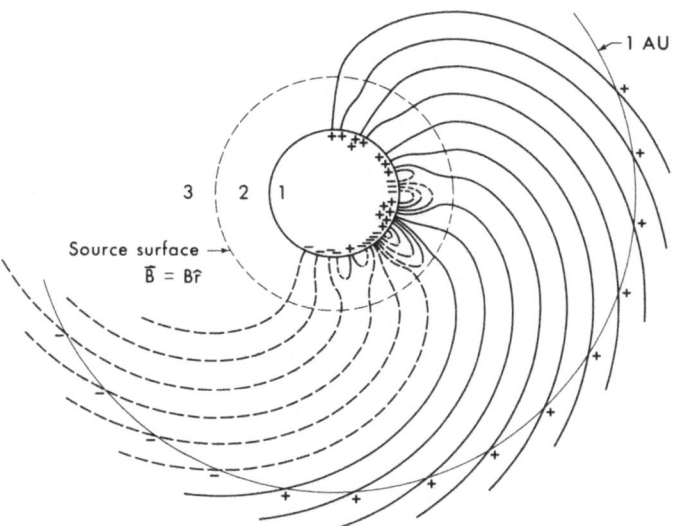

1. Photospheric magnetic field observed at Mount Wilson
2. Magnetic field calculated from potential theory $\nabla^2 \phi = 0$
3. Magnetic field transported by solar wind $\dfrac{d\bar{B}}{dt} = -\bar{B}(\nabla \cdot \bar{V}) + (\bar{B} \cdot \nabla)\bar{V}$
   (observed by spacecraft at 1 AU)

Fig. 5.21   A sketch illustrating the computation of the coronal magnetic field configuration in the "source surface" model [5.32]

realistic B_r (r_0, ϕ, θ). A workable boundary condition at $r = r_s$, the specification that the magnetic potential be constant or the field be purely radial at this surface, has been extensively employed by Schatten, Wilcox, and Ness [5.32]. The entire distribution of currents implicit in the detailed plasma-field interaction is effectively replaced by a sheet current at $r = r_s$.

This so-called "source surface model" is illustrated schematically in Fig. 5.21. Region 1 represents the photosphere, with a complex magnetic configuration illustrated by the plus and minus signs. Region 2 is the spherical shell $r_\odot < r < r_s$ wherein the field is computed from a scalar potential. Within this region, closed field lines would be expected, in particular above photospheric regions of complex magnetic polarity. After becoming radial at the "source surface" $r = r_s$, the field lines of region 3 follow the spiral configuration implied by an overall radial expansion. No field lines can close in Region 3.

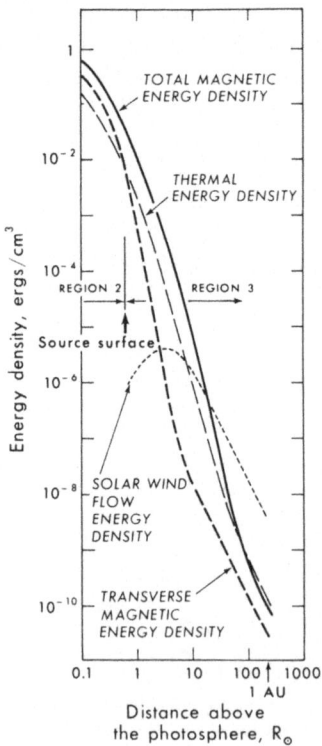

Fig. 5.22 Estimated magnetic and plasma energy densities in the corona and interplanetary space [5.32]

Any quantitative application of the source surface model outlined above requires specification of the radius r_s. A logical choice of r_s would be the heliocentric position at which the energy in the transverse magnetic field (the component capable of affecting the outward expansion) becomes less than the energy in the plasma. Fig. 5.22 shows estimated energy densities in the coronal magnetic field and plasma as functions of distance above the photosphere [5.32]. On the basis of the criterion given above, $r_s \approx 1.6 r_\odot$. Schatten, Wilcox, and Ness found that this choice also produced reasonable agreement between the average field magnitude and frequency spectrum computed at the source surface and that observed near 1 AU. Altschuler and Newkirk [5.33] found that a different choice, $r_s = 2.5 r_\odot$, led to the best match between the calculated field configuration and the coronal structure observed during the solar eclipse of November, 1966.

The accuracy with which a source surface model approximates the coronal magnetic field configuration can be tested by comparison with the accurate solution for the dipole case, already described above. Fig. 5.23 compares the field lines computed by Pneuman and Kopp [5.31] with those computed for a dipole at $r = r_\odot$ and a source surface at $r = 2.49 r_\odot$ [5.34]. Agreement is quite good, especially near the region of closed field lines. For this particular case, it is clear that the source surface model (with careful choice of the radius r_s), despite its approximate treatment of the plasma-field interaction, does give a reasonable representation

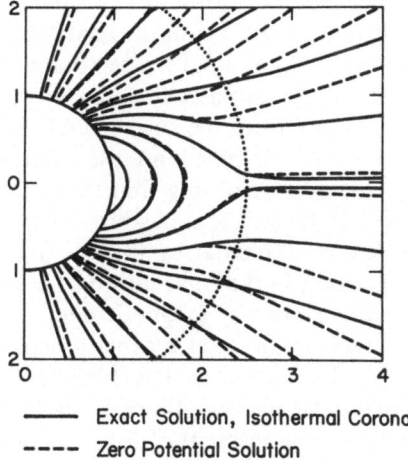

——— Exact Solution, Isothermal Corona

– – – – Zero Potential Solution

Fig. 5.23   The magnetic field configuration predicted by the detailed coronal expansion model of Pneuman and Kopp (solid lines) and the source surface model (dashed lines) with a dipole field at $r = r_\odot$ [5.34]

of the magnetic field configuration. The advantage of this model then becomes its flexibility; it can be applied to complex and realistic fields at $r = r_\odot$ with hope of obtaining a reasonable description of the coronal fields, but without requiring solution of the coupled fluid equations and Ampere's law. The disadvantage is that the model gives no information regarding the plasma flow characteristics along the resulting open field lines.

The actual application of the source surface model to the computation of coronal and interplanetary field characteristics will be illustrated for the period July 30 to Aug. 25, 1965. Fig. 5.24 shows a synoptic map of the photospheric magnetic field determined by the Mt. Wilson solar magnetograph [5.35]. The dark gray regions represent fields observed to be pointing into the sun, while the light gray regions represent fields observed to be pointing out of the sun. Contours of the photospheric field identity are shown at 6, 12, 20, and 30 gauss by light lines within the shaded regions. The heavy lines are contours at 0.25 and 0.75 gauss of the magnetic field computed at the source surface, $r = 1.6\,r_\odot$; solid source-surface contours represent fields pointing away from the sun, while dashed source-surface contours represent fields pointing toward the sun [5.36]. The source surface field configuration clearly reflects the overall characteristics of the photospheric configuration, but lacks many of the latter's complexities. For example, the small region of outward directed photospheric field near 250° in the northern solar hemisphere does not produce an outward field region on the source surface. The interplanetary magnetic polarity observed on the Imp 3 satellite is indicated along the bottom of the synoptic chart (sectors with field pointing away from the sun as light shading, sectors with field pointing toward the sun as dark shading) with a shift of five days to approximate the transit of a magnetic feature from the sun to 1 AU (V.3). The source surface polarity near $\sim 30°\,N$ latitude and the interplanetary sector pattern are in good agreement, suggesting that the magnetic field from this coronal region (near 30° N latitude) has been convected into interplanetary space with little further change in its basic topology. The large regions of predominant polarity in the solar photosphere again emerge as the likely sources of the interplanetary magnetic field.

The relationship between high-speed plasma streams and magnetic sectors discussed in V.4 then implies that high-speed streams also originate within such magnetic regions. Since solar wind observations have been published for the time interval of Fig. 5.24, let us pursue this implication in more detail. Fig. 5.25 shows three-hour averages of the solar wind flow speed, proton density, and proton temperature observed by Vela 3 spacecraft and the magnetic field intensity and polarity observed by the Imp 3 spacecraft between July 15 and Sept. 6, 1965 [5.5].

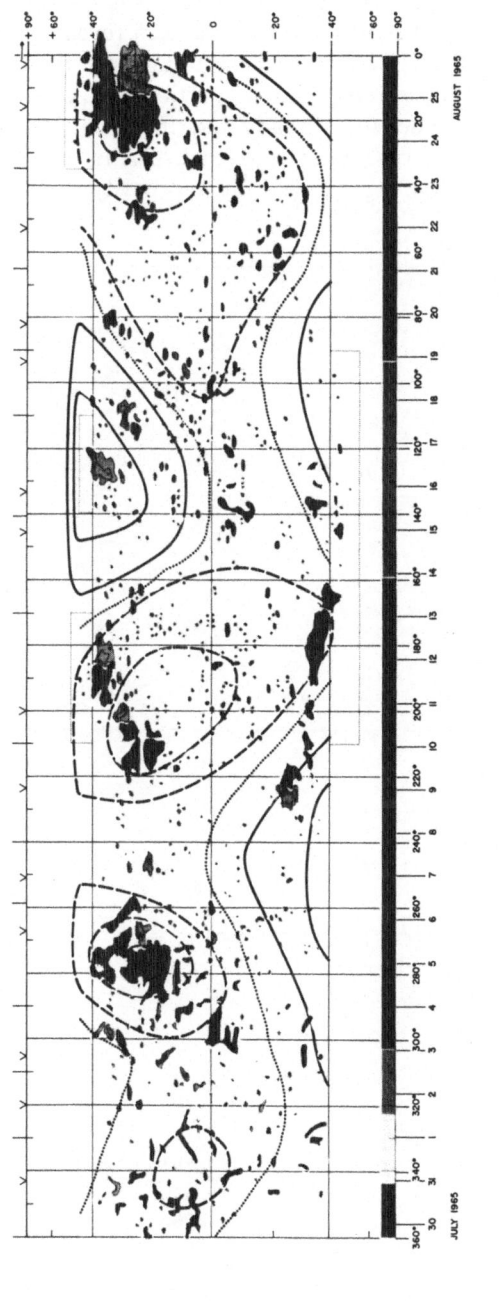

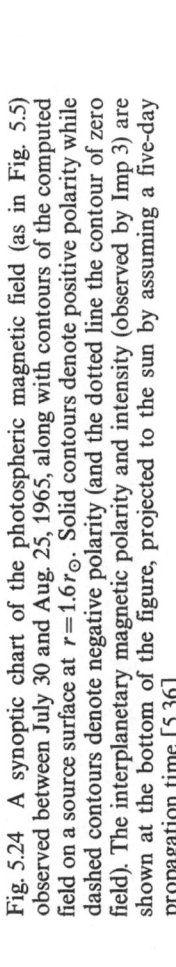

Fig. 5.24 A synoptic chart of the photospheric magnetic field (as in Fig. 5.5) observed between July 30 and Aug. 25, 1965, along with contours of the computed field on a source surface at $r = 1.6 r_\odot$. Solid contours denote positive polarity while dashed contours denote negative polarity (and the dotted line the contour of zero field). The interplanetary magnetic polarity and intensity (observed by Imp 3) are shown at the bottom of the figure, projected to the sun by assuming a five-day propagation time [5.36]

The magnetic sector pattern (basically that already used in Fig. 5.24) was well defined during this interval, and inspection of Fig. 5.25 reveals that a high-speed plasma stream existed within each sector. These plasma streams displayed the general variations, and in particular the compression-rarefaction pattern in the density, already described in V.2. One well-defined, but small, high-speed plasma stream, that of Aug. 14, was also observed in the midst of a large region of negative polarity.

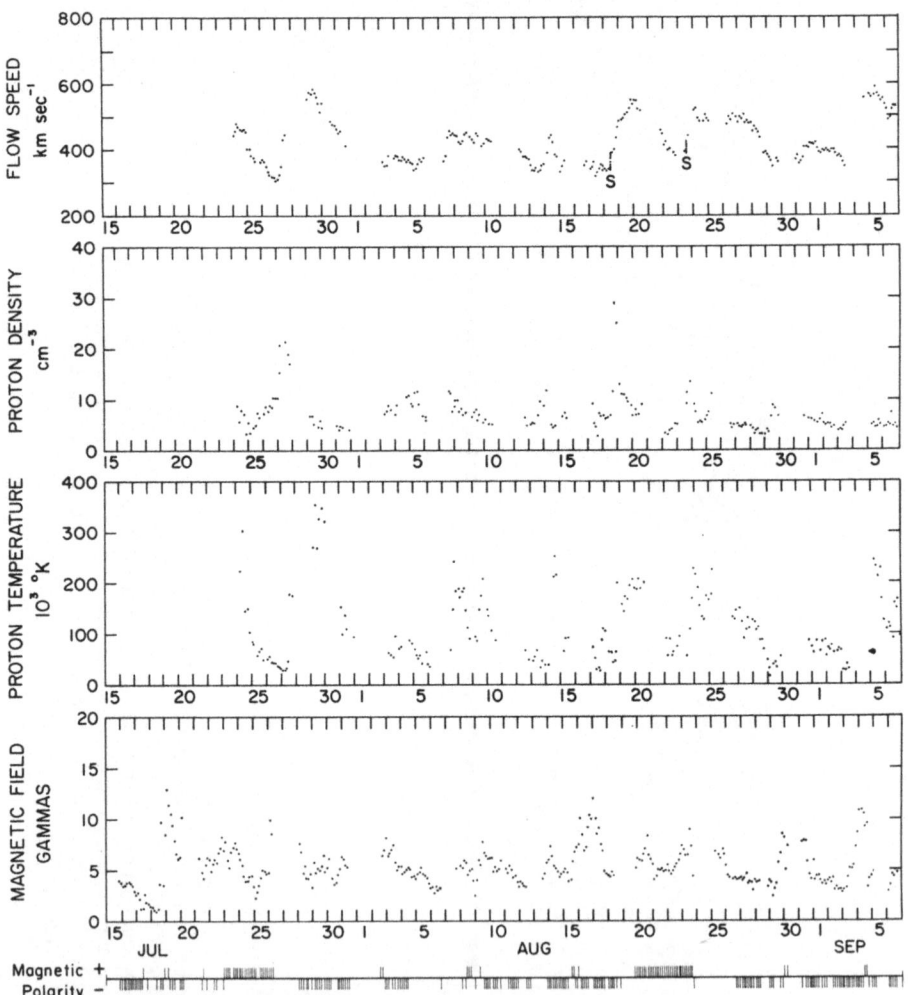

Fig. 5.25   Solar wind plasma and magnetic field properties observed by the Vela 3 and Imp 3 spacecraft between July 15 and Sept. 6, 1965 [5.5]

Fig. 5.26 combines a variety of solar and interplanetary observations with results of a source surface computation for the interval July 30 to Aug. 25, 1965. The computed field lines are shown at 30° N latitude for the range $r_\odot \leq r \leq r_s = 1.6 r_\odot$, with solid lines indicating outward-pointing fields, dashed lines indicating inward-pointing fields [5.36]. The formation of closed field lines above complex magnetic regions is clearly displayed. In particular, the small region of inward-directed fields that was at central meridian on Aug. 8 is connected to nearby regions of outward field by low-lying magnetic loops; hence its already-mentioned failure to produce an inward-pointing field on the source surface. The daily average value of the radial magnetic field component

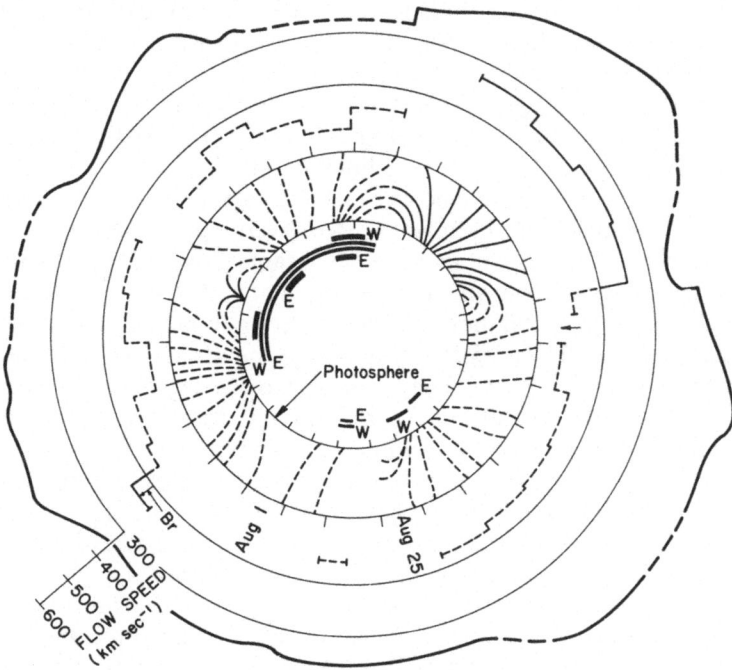

Fig. 5.26   Solar and interplanetary data from the interval July 30 to Aug. 25, 1965. The coronal green line observations (thin line for moderate emission, thick line for bright emission) at 30° N latitude are shown within the circle denoting the photosphere. The field configuration computed at the same latitude from a source surface model is shown for $r_\odot \leq r \leq 1.6 r_\odot$ [5.36]. The daily average radial component of the interplanetary magnetic field (observed by Imp 3) is shown surrounding the field map; these observations have been advanced by five days to account for transit from the sun to 1 AU. The Vela 3 flow speed observations of Fig. 5.25 are shown around the outermost circle; these observations have been advanced by the transit time appropriate to the observed expansion speed at 1 AU

observed by Imp 3 is shown on the circle surrounding the field configuration, with solid lines again indicating outward-pointing field,
dashed lines indicating inward-pointing field. This interplanetary
polarity has been advanced by five days to compensate for the expected
transit time of the frozen-in field from the sun to 1 AU. The correspondence
between the source surface and solar wind polarity pattern is encouraging-
ly close. The Vela 3 flow speed observations of ~Aug 4 to Aug 29 from
Fig. 5.24 are represented by a continuous curve (dashed where extra-
polated through an absence of actual data) along the outermost circle of
Fig. 5.25. Each portion of this curve has been advanced in time by r_e/u,
where u is the actual observed flow speed, to compensate for the transit
time from the sun to $r = r_e = 1 AU$. For example, the 550 km sec^{-1}
plasma observed on Aug. 20 is shown above the photospheric field that
passed central meridian on Aug. 16, or 3.2 days earlier. This nonuniform
advancement of the flow speed measurements distorts the observed
profile but retains the general pattern of high-speed streams. The resulting
flow speed vs longitude curve on Fig. 5.26 generally shows low expansion
speeds over regions of closed field lines, and high expansion speeds over
the diverging field lines connected to large regions with a predominant
polarity. This rule holds even for the two diverging field regions of Aug. 4
and 10, despite the fact that they are of the same polarity. A low expansion
speed was observed over the intervening region of closed field lines, even
though this region did not produce a sector boundary. It should be
emphasized that the "projection" of interplanetary observations back to
the sun by invoking a time delay is an approximate procedure, based on
the questionable assumption of a constant, radial expansion speed.
Nonetheless, the concept of channeling of the expanding plasma by the
coronal magnetic field, producing high-speed plasma streams with an
organized magnetic polarity, is supported by this example.

Fig. 5.26 also shows regions of coronal green line emission at 30° N
latitude, along the inside of the circle representing the photosphere.
The thin lines denote moderate emission, while the thicker lines denote
strong emission. The symbols E and W signify that this emission was
observed when the region was on the east or west limb of the visible
sun. The green line emission is produced by the ion Fe^{+13}, and indicates
a moderate elevation of the coronal temperature, generally associated
with regions of moderate solar activity. Although it might be tempting
to attribute the presence of high-speed plasma streams to active regions,
Fig. 5.26 lends little support to this idea. The strongest green line emission
appears to occur in regions with closed field lines. The presence of green
line emission in the unipolar region near the Aug. 10 position does not
produce a high-speed stream of unusual magnitude or duration (even
though the observation of the emission on both the east and west limbs

indicates that the associated temperature elevation was long-lived). Inso-far as the green line is an index of solar activity or enhanced coronal temperature, this example implies little relationship between these con-ditions and the solar sources of high-speed plasma streams.

On the basis of this example, we are led to the tentative conclusion that high-speed plasma streams emanate from the coronal regions of diverging magnetic fields, above the large regions with a predominant magnetic polarity that are observed in the photosphere. This empirically-based concept is qualitatively consistent with the dipolar latitude varia-tion in expansion speed predicted by the model of Pneuman and Kopp. It is intermediate between the pure "nozzle" and "mapping" hypotheses (V.3), but closer to the latter; Schatten [5.36] has estimated that $\sim 1/3$ of the photosphere is mapped onto the source surface at $r = 1.6 r_\odot$. The discussion at the end of this chapter (V.9) will attempt to place our tentative identification of the source of high-speed streams in a more general context.

V.7 The Effects of High-Speed Plasma Streams on Average Solar Wind Characteristics

The presence of high-speed plasma streams, producing large variations of all plasma properties except for the electron temperature, must affect solar wind characteristics deduced from long-term statistical analyses of interplanetary observations. The most obvious effect would be on long-term averages; for example, the mean flow speed and proton temperature would be elevated above values relevant to a hypothetical steady, spheri-cally-symmetric background expansion if high-speed streams were included in an observational sample. Our attempt (III.2 and III.3) to carefully define an observed structureless solar wind state was based on this realization. More subtle effects are possible when statistical relation-ships among various solar wind properties are derived from a body of solar wind observations.

It has become almost customary to regard the solar wind speed as an "independent" parameter for analyses of observations, and to search for relationships between the flow speed and other "dependent" para-meters. This point of view clearly stems from the use of the flow speed as a criterion for quiet or disturbed conditions, as discussed in III.2. The "dependence" of the density, flow direction, proton and electron tem-perature, magnetic field intensity and direction upon the flow speed have all been discussed in this context [5.1, 5.2, 5.3, 5.37, 5.38, 5.39, 5.40]. For example, Fig. 5.27 shows the average proton density observed by Vela 3 spacecraft when the flow speed, u, was in the $25\,\mathrm{km\,sec^{-1}}$ ranges

$250 \le u < 275 \, \text{km sec}^{-1}$, $275 \le u < 300 \, \text{km sec}^{-1}$, etc. [5.39]. The average density depends inversely on the flow speed, decreasing from $\sim 9 \, \text{cm}^{-3}$

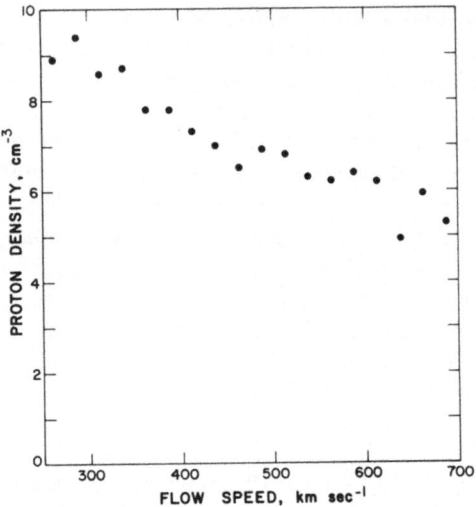

Fig. 5.27   The average proton density observed in $25 \, \text{km sec}^{-1}$ flow speed intervals by the Vela 3 spacecraft [5.39]

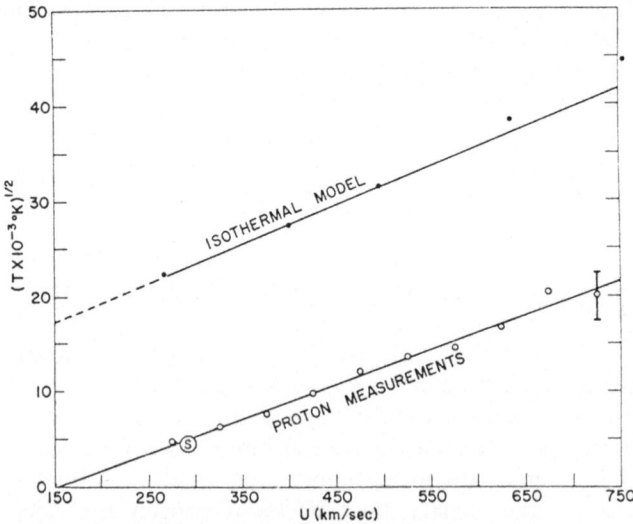

Fig. 5.28   The proton temperatures observed in $50 \, \text{km sec}^{-1}$ flow speed intervals (open circles) by the Explorer 34 spacecraft [5.2]. The straight line through the observations is the empirical relation suggested by Burlaga and Ogilvie (see text). The predictions of Parker's isothermal model (Fig. 1.2) are shown for comparison

for $u=300\,\mathrm{km\,sec^{-1}}$ to $\sim 6\,\mathrm{cm^{-3}}$ for $u=700\,\mathrm{km\,sec^{-1}}$. Fig. 5.28 shows the square root of the average proton temperature observed on the Explorer 34 spacecraft when the flow speed was in the $50\,\mathrm{km\,sec^{-1}}$ ranges $250\le u<300\,\mathrm{km\,sec^{-1}}$, $300\le u<350\,\mathrm{km\,sec^{-1}}$, etc. [5.2]. The proton temperature depends directly on the flow speed, rising from $\sim 2.5\times10^4\,^\circ\mathrm{K}$ for $u=300\,\mathrm{km\,sec^{-1}}$ to $\sim 4\times10^5\,^\circ\mathrm{K}$ for $u=700\,\mathrm{km\,sec^{-1}}$. The straight line on Fig. 5.28 indicates an empirical representation of this relationship proposed by Burlaga and Ogilvie [5.2];

$$\sqrt{\frac{T_p}{10^3}}\,^\circ\mathrm{K}=0.036\,u-5.44\,,\qquad(5.11)$$

where u is in $\mathrm{km\,sec^{-1}}$. Fig. 5.29 shows the average flow direction observed by the Vela 2 satellites for the same $25\,\mathrm{km\,sec^{-1}}$ flow speed intervals used in the density-flow speed example above [5.38]. The solar wind flows from farthest east of the sun (Fig. 3.20) at low flow speeds, shifting westward with increasing flow speed.

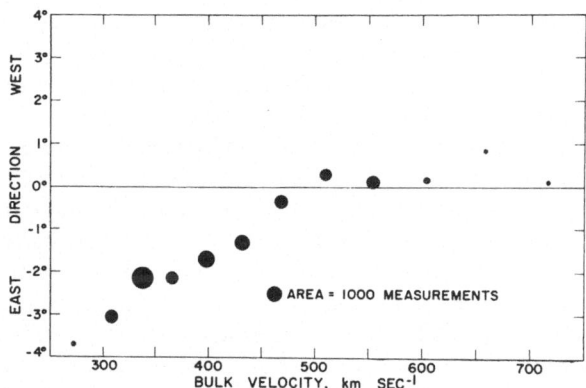

Fig. 5.29 The average flow direction observed in 25 km sec^{-1} flow speed intervals by Vela 2 spacecraft [5.38]

A complete theoretical understanding of the coronal expansion should provide physical explanations of these relationships. Some attempts in this direction have been based on models assuming a steady, spherically-symmetric coronal expansion. For example, it has been found that proper choice of the parameters in the extended energy source term assumed in the Hartle and Barnes modification of the basic two-fluid solar wind model (III.8) leads to a sequence of models whose predicted flow speeds and temperatures at 1 AU match those in equation (5.11) [5.41]. Similarly, variation of the hydromagnetic wave flux at $r=2r_\odot$ in the model of Barnes, Hartle, and Bredekamp [5.42] leads to a sequence of models in rough agreement with (5.11). This agreement is taken as evidence that high-speed solar wind streams are produced by an intensification of the extended energy deposition invoked in these models to explain detailed properties of the structureless solar wind.

In view of the long discussion of the time scales of solar wind variations and the implications regarding the applicability of steady, spherically symmetric models to the interpretation of these variations (II.2, II.3, III.2), it should come as no surprise to the reader that the author regards the arguments presented in the preceding paragraph with considerable skepticism. In the light of the assembled observational evidence that variations in solar wind properties are dominated by *dynamic* processes, with scale times comparable to that of the basic coronal expansion, the burden of proof for the applicability of steady, spherically-symmetric models to the interpretation of interparameter relations, inferred from an undifferentiated body of observations, must fall squarely upon the advocates of such models. No valid proofs of this nature have been given.

In contrast, it is clear that the known relationships among the solar wind speed and the density, flow direction, proton temperature, and electron temperature, are all qualitatively consistent with the observed pattern of *correlated* variations in high-speed plasma streams (V.2) and with the variations predicted by the theoretical models of high-speed streams (V.5). The flow direction variations have, in fact, been interpreted in terms of an inhomogeneous flow model for some time [5.4]. The development of such models adequate to give quantitative explanations of the observed relationships is inherently more difficult than the formulation of steady, spherically-symmetric models. Nonetheless, even the approximation of a high-speed stream as a transient disturbance, as shown in Figs. 5.15 and 5.16 leads to low densities and high temperatures at high flow speeds. Fig. 5.30 shows the relationship between the (time-averaged) proton temperature and the flow speed for the high-speed stream model used in Fig. 5.16 (including a similar computation resulting in a higher speed) [5.43]. The average proton temperatures observed in $25 \, \mathrm{km \, sec^{-1}}$ flow speed intervals by the Vela 3 satellites are also shown for comparison. The high-speed stream model implies a relationship between temperature and flow speed quite similar to that actually observed. In fact, the time-averaged temperature structure of these streams is not very different from that predicted by a steady, adiabatic model valid over the same 0.15 to 1 AU range of heliocentric distance. The significant difference between the steady and "dynamic" predictions is that in the latter the highest temperatures occur on the rising portion of the velocity variation. Fig. 5.30 demonstrates that the observed relationship between T_p and u is consistent with a dynamic interpretation, and removes any necessity for invoking a particular source of the added energy in the stream, as assumed in the extended heating models.

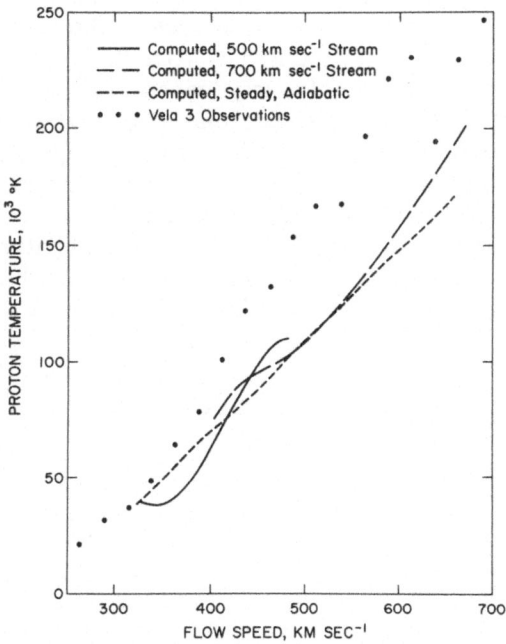

Fig. 5.30 The relationship between temperature and flow speed predicted by two cases of transient high-speed streams (as in Fig. 5.16), by a steady adiabatic model for $r > 0.13 r_e$, and observed by the Vela 3 spacecraft [5.43]

V.8 Energy Transport in High-Speed Solar Wind Streams

The elevated densities and flow speeds observed near the leading edges of high-speed plasma streams indicate an enhanced rate of energy transport from the corona. We are thus led back to another consideration of the problem of solar wind energetics. Two basic questions arise: Is an extended energy source, as employed in some models of the steady, spherically-symmetric coronal expansion (III.7 and III.8), required to produce the high expansion speeds characteristic of the streams? How important are high-speed streams in the overall flow of energy from the solar corona?

The reader will recall that no firm conclusions emerged from our extensive discussion of the role played by extended heating in the structureless coronal expansion, and should then not be surprised to learn that the energetics of high-speed streams are still more poorly understood. The arguments for extended heating in the streams are similar to those already applied to the structureless solar wind. Parker [5.44] pointed out that the high expansion speeds observed in streams, say $600 \, \text{km} \, \text{sec}^{-1}$, could not be produced by a one-fluid, conductive expansion model without assumption of a very low coronal density ($\sim 10^7 \, \text{cm}^{-3}$). Hartle and Sturrock [5.45] argued that a solar wind speed of $750 \, \text{km} \, \text{sec}^{-1}$ would require an unrealistic coronal temperature, as high as $7.7 \times 10^6 \, ^\circ \text{K}$ in the basic two-fluid model, and also lead to a solar wind density of $\sim 10^3 \, \text{cm}^{-3}$, far higher than actually observed. Thus an extended energy source, probably related to dissipation of the energy carried by waves, was again postulated to increase the energy in (and speed of) the coronal expansion. While these arguments are reasonable,

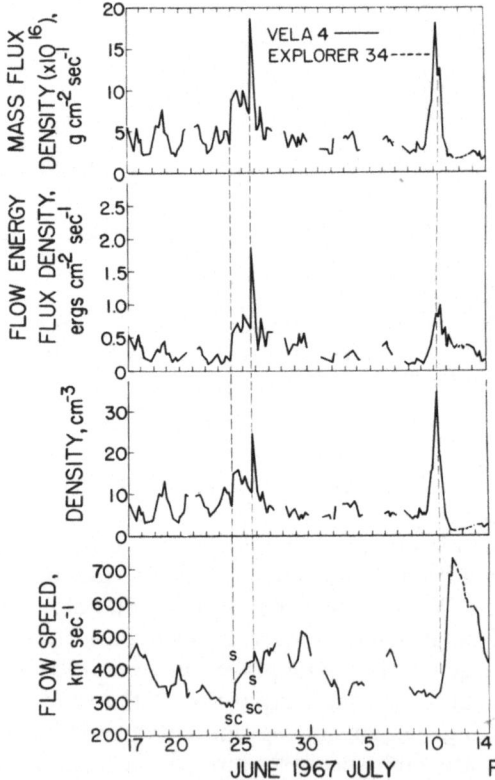

Fig. 5.31 Vela 4 and Explorer 34 observations of proton density, flow speed, mass flux density, and energy flux density from a single solar rotation in mid-1967 [5.46]

they are by no means compelling. The application of models assuming a steady, spherically-symmetric flow to the complexities of high-speed streams remains highly questionable. The nonlinear evolution of a stream in transit from the sun to 1 AU (V.5) can obscure the original density structure. Magnetic channeling of the expansion in the low corona (V.6) could lead to a ratio of coronal and interplanetary densities far different from that for a spherically-symmetric expansion. Thus the basic necessity for extended heating remains unclear.

Little quantitative observational information on the energetics of high-speed streams has become available until very recently. Fig. 5.31 shows Vela 4 and Explorer 34 observations of the proton density and flow speed for a single 27-day solar rotation in mid-1967 [5.46]. To these basic fluid parameters have been added the mass flux density mnu and the energy flux density $\frac{1}{2}mnu^3$. A large high-speed stream appeared on July 11, with a rise in flow speed from $\sim 320\,\mathrm{km\,sec^{-1}}$ to over 700 km sec^{-1} in one day, followed by the typical slow decline. Extremely high densities (~ 7 times the apparent prestream level) were observed near the leading edge of the stream; however the high densities persisted for only ~ 1 day, and were followed by a particularly striking rarefaction. The energy and mass flux densities were very high only during the large compression, dropping to moderate values in the rarefaction despite the persistence of high flow speeds. Thus the enhanced mass and energy fluxes associated with the stream were concentrated at its leading edge. This is, of course, another manifestation of the "steepened" nature of the high-speed streams observed in interplanetary space.

Estimation of the energy flux in this stream requires some assumption regarding the area over which the flux density is enhanced. We will

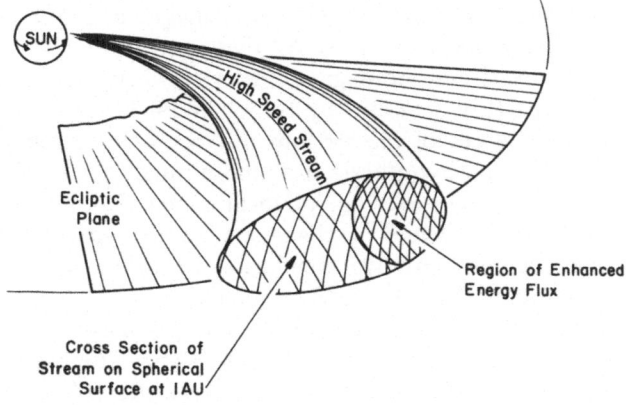

Fig. 5.32   The geometry assumed in estimating the energy flux in a high speed solar wind stream

assume equal latitude and longitude extents for this area, as illustrated in Fig. 5.32, and infer the latter from the temporal width of the observed flux density peak. For the example under discussion, the resulting estimate of the flux through a spherical surface at $1\,\mathrm{AU}$ is $\sim 2 \times 10^{25}\,\mathrm{ergs\,sec^{-1}}$. By way of comparison, the energy flux density in the low-speed solar wind (Table 3.2) implies a total energy flux through the entire sphere at $1\,\mathrm{AU}$ of $6 \times 10^{26}\,\mathrm{ergs\,sec^{-1}}$. Despite the large enhancement of the energy flux density near the leading edge of this prominent high-speed stream, the short duration of that enhancement leads to a total flux that is only a few percent of that in the quiet solar wind.

The example given above is entirely typical. Analysis of Vela 3 and 4 data from ten solar rotations during the interval Oct. 1965 to Dec. 1967 reveals that the combined energy flux in all observed high-speed streams (assuming that they persist for the entire solar rotation) averages about 20% of the flux associated with low-speed solar wind [5.46]. The implications with regard to the two questions posed earlier may be somewhat unexpected. The observations do not indicate that a large energy flux is required to produce high-speed solar wind streams. Although the energy flux in the streams is by no means negligible, it never appears to dominate the overall efflux of energy in the coronal expansion.

V.9 The Solar Sources of High-Speed Plasma Streams

Our earlier discussion of models treating the interaction of plasma and magnetic fields in the corona (V. 6) led to a tentative conclusion that the large photospheric regions with predominant magnetic polarities (and the diverging magnetic field configurations above them) are plausible solar sources of the interplanetary magnetic sectors and associated high-speed plasma streams. As these interplanetary structures were related to recurrent geomagnetic activity when first observed by Mariner 2 in 1962 (Fig. 5.6) and by Imp 1 in 1963 (Fig. 5.4), it would appear that the sources of the streams and sectors are related to the M-regions [5.47], long hypothesized as the ultimate cause of recurrent geomagnetic activity but never unequivocally identified with any visible solar feature. Let us conclude the discussion of high-speed streams and sectors by placing our tentative conclusion regarding their origins in the general context of the "M-region problem."

Historically, two distinct schools of thought have evolved from attempts to identify the M-regions. One of these has attempted to relate the recurrent geomagnetic activity to specific centers of solar activity. The bright streamers observed during solar eclipses have often been suggested as coronal structures indicating the presence of localized emission of material from active regions [5.48]. The second school has attempted to relate the recurrent geomagnetic activity with the "quiet" regions of the sun, arguing for a "cone of avoidance" of solar corpuscular emission above active regions [5.49, 5.50]. Billings and Roberts [5.51] have suggested that the coronal structure over a relatively stable active region resembles that shown in Fig. 5.33. In a conducting corona, both the non-radial and strongly divergent nature of the field lines near the active region imply the depression of isotherms shown on the figure. In the context of Parker's coronal expansion models [5.44], the steeper radial temperature gradient in turn implies emission of low-speed solar wind above the active region.

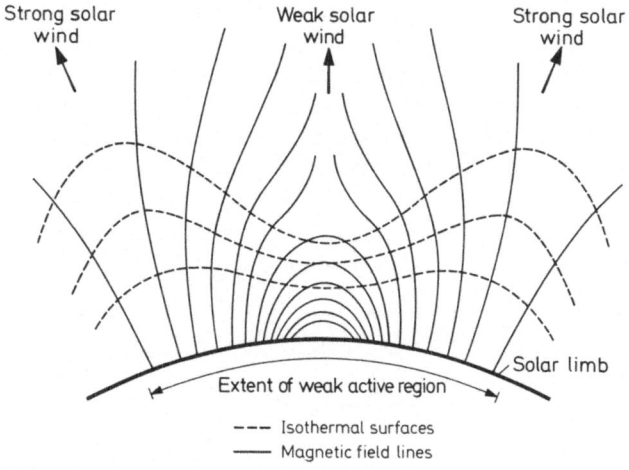

Fig. 5.33 The coronal magnetic and thermal structure proposed to explain the "cone of avoidance" by Billings and Roberts [5.51]

Several attempts have been made to associate the high-speed streams observed in the solar wind with specific solar features. Snyder and Neugebauer [5.52] projected the interplanetary features observed by Mariner 2 back onto the solar disk by assuming radial expansion at the measured solar wind speed (as in the construction of Fig. 5.26) and compared the inferred sources of the high-speed streams with the locations of observed calcium plages. No convincing relationship was found. More recently, Couturier and Leblanc [5.53] have claimed a close relationship between the projected solar locations of the Mariner 2 high-speed streams and coronal enhancements, observed in both metric radio emission and coronal green line emission. These enhancements, both denser and hotter than the surrounding corona, are regarded as the sources of the high-speed streams through formation of a con-stricted, "streamer-like" local expansion above each enhancement. Yet, some aspects of this study are not entirely convincing. Although the numbers of coronal enhancements and high-speed streams observed during each solar rotation are nearly the same, more of the enhance-ments appear to coincide with the projected positions of flow speed minima than with the projected positions of the streams. Further, it is not clear that a supersonic expansion is possible in streamers with the high densities ($\sim 7.5 \times 10^7\,\text{cm}^{-3}$ at $r = 2r_\odot$) indicated by the radio observations of the coronal enhancements [5.30].

A most important objection to the general association of high-speed solar streams with regions of solar acitivity is the surprising absence of any major changes in average solar wind properties observed during the rising portion of the present solar cycle. Gosling, Hansen, and Bame [5.54] combined the solar wind speed observations made by numerous spacecraft between 1962 and 1969. Despite the great change in solar activity during this interval (including a twofold increase in the white light coronal intensity), no large trend in the flow speed was discernible.

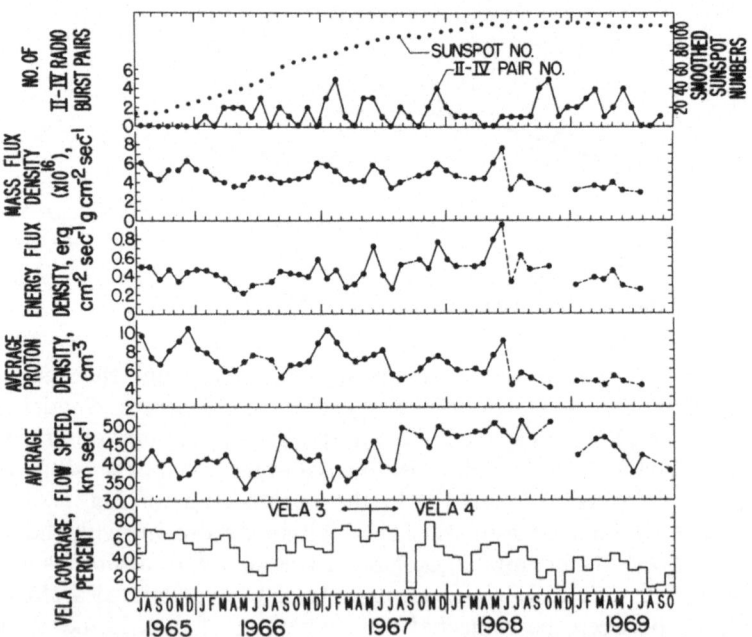

Fig. 5.34 Twenty-seven day averages of solar wind properties observed by Vela 3 and Vela 4 spacecraft between 1965 and 1969, along with two indices of solar activity [5.46]

Fig. 5.34 shows 27-day averages of the solar wind properties (including the mass and energy flux densities) observed by Vela 3 and Vela 4 spacecraft between 1965 and 1969 [5.46], along with several indices of solar activity. The lack of any apparent relationship between the rising level of solar activity and the nearly constant values of solar wind properties would seem to argue against regions of solar activity as the sources of any *common* solar wind phenomena.

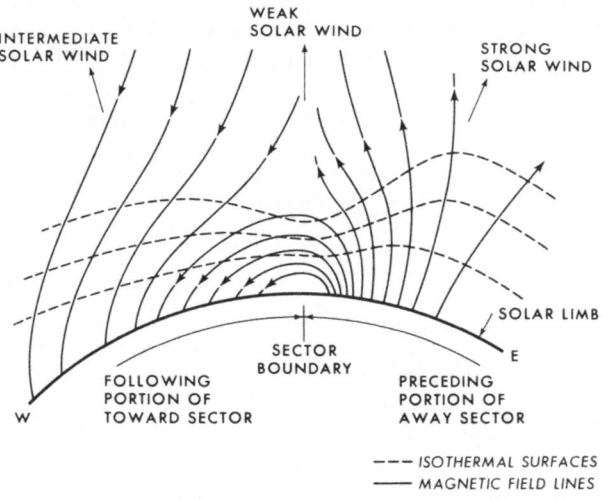

Fig. 5.35   The coronal magnetic and thermal structure near a solar sector boundary, as proposed by Wilcox [5.12]

The widely accepted identification of the large photospheric regions with predominant magnetic polarities as the sources of interplanetary sectors [5.10, 5.11, 5.12] is in general accord with the basic tenets of the "cone of avoidance" school. Fig. 5.35 shows an adaption by Wilcox [5.12] of Fig. 5.33 to represent the coronal structure near the boundary of two unipolar regions on the sun ("solar sectors"). An expansion speed variation, similar to that of Billings and Roberts, is indicated; however, a distinction between the flow on the two sides of the boundary has been postulated to explain the occurrence of the highest flow speeds in the leading portion of interplanetary sectors. A similar configuration has been recently proposed by Sakurai and Stone [5.55], who have found centers of type I radio emission near the solar sector boundaries and argued that this radio emission originated on the closed field lines above active regions.

Our tentative identification of the coronal regions of diverging magnetic field lines above such photospheric regions as the sources of high-speed plasma streams is consistent with the ideas of Wilcox and of Sakurai and Stone, and again fits into the "cone of avoidance" school. It should be emphasized that this identification is based upon the best available *physical* models of magnetic channeling in the expanding corona as well as upon the somewhat unsure process of projecting observed interplanetary features back to the sun. In fact, combination of the hydromagnetic expansion model of Pneuman and Kopp (valid only near the sun because of its isothermal nature and neglect of solar rotation) and the nonlinear stream interaction models of V. 5 (valid only

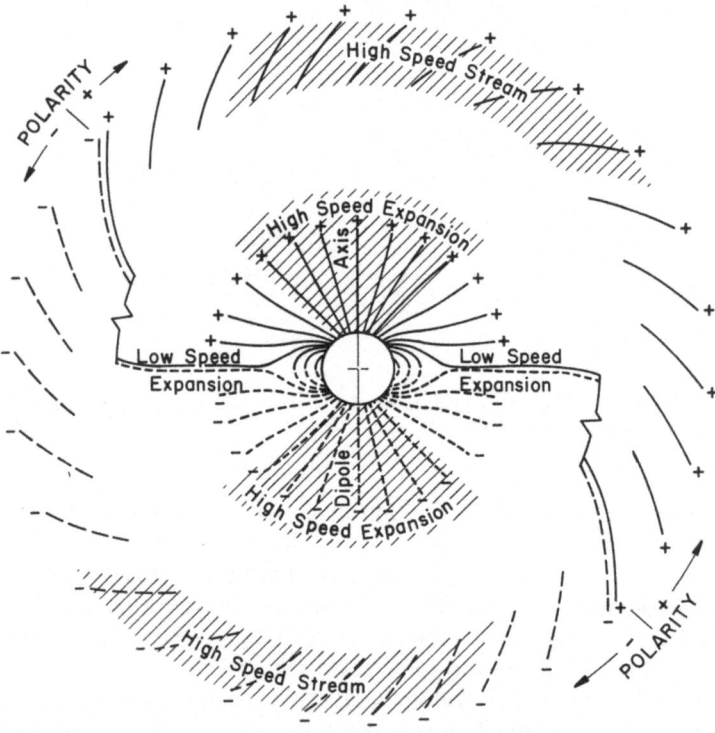

CROSS SECTION IN SOLAR EQUATORIAL PLANE

Fig. 5.36   A schematic combination of the coronal expansion model of Pneuman and Kopp (Figs. 5.19 and 5.20) near the sun (with the dipole axis assumed to lie in the solar equatorial plane) with a nonlinear stream model (such as Figs. 5.14 or 5.16) at larger heliocentric distances. The higher expansion speed near each dipole axis leads to two high-speed solar wind streams, one within each magnetic unipolar region or sector. The high-speed streams have moved into the leading portions of each sector in propagating from the sun to a large heliocentric distance

well away from the sun because of the neglect of magnetic channeling) can predict a simple pattern of interplanetary plasma and magnetic variations basically similar to that actually observed. Fig. 5.36 shows the Pneuman and Kopp model with the dipole axis assumed to lie in the solar equator. Two longitude regions of uniform magnetic polarity result. The expansion speed is higher near the dipole axis than at the longitudes over the closed field line regions at the dipolar equator (Fig. 5.20). This longitude variation in expansion speed would presumably be enhanced in a model including heat conduction (as indicated on Figs. 5.33 and 5.35, whose resemblance to 5.36 should be obvious). The subsequent modification of this longitude variation by nonlinear steepening and by solar rotation should be very similar to that illustrated on Fig. 5.14. The expected structure near 1 AU is sketched along the outside of Fig. 5.36. Two magnetic sectors, each containing a high-speed stream with an asymmetric profile, would be observed in interplanetary space.

It would be misleading to overemphasize the certainty with which our conclusion regarding the source of high-speed streams can be drawn from our limited present-day understanding of coronal and interplanetary structure. Nonetheless, it is this author's opinion that the point of view presented above, in general accord with the "cone of avoidance" hypothesis, is that best supported by solar wind observations and theory. Note that this point of view emphasizes the variation of expansion speed across a unipolar region of diverging coronal field lines. This structure would presumably be present regardless of any polarity difference with respect to neighboring source regions (as, for example, at the doubly-closed field line region near Aug. 10 on Fig. 5.26). Thus the streams are regarded as more fundamental structures than the magnetic sectors. We emphasize again that no asymmetry in this variation across the source region is *required* to account for the asymmetric velocity profile observed in interplanetary space if the nonlinear evolution of the resulting wave is recognized.

Even if this conclusion regarding the *spatial* source of high-speed plasma streams is accepted, several possibilities remain for the *energy* source of the streams. The entire variation in expansion speed shown in Figs. 5.20 and 5.36 is due to the magnetic forces on the plasma. The effect of the field on the temperature gradient in a conducting corona (Figs. 5.33 and 5.35) has been verified by direct integration of the hear transfer equation [5.56], so that a further enhancement of this expansion speed variation could be expected. As both the energy and mass problems will assume new aspects in the diverging flow pattern suggested here (e.g., Parker [5.44]), any consideration of the adequacy of the conduction corona or the necessity for an extended energy source should be based on more sophisticated models than are presently available. It is not even clear that coronal observations have determined the physical conditions pertinent to the proposed source regions. It is possible that the coronal heating mechanism (in either a "thin shell" or "extended" form) is related to the coronal magnetic structure and will thus vary with position across a "magnetically-organized" source region. This hypothesis may be suggested by an oberved concentration of interplanetary Alfvén waves (II.3 and III.8) in and near the leading edges of high-speed streams [5.57]. It should be abundantly clear that our physical understanding of the nonuniform coronal expansion and the origin of high-speed streams and magnetic sectors, although imperfect, has suggested some interesting possibilities.

Chapter VI

Flare-Produced Interplanetary Shock Waves

VI.1 Introduction

The occurrence of geomagnetic storms one to several days after some prominent solar flares has long been attributed to arrival at the earth of material ejected by the flare. Consider a volume of this material, presumably plasma from the chromosphere or low corona, moving rapidly outward into a slower ambient solar wind (such as already shown in Fig. 5.9). Fig. 6.1 shows a hypothetical cross section (in the solar equatorial plane) of the resulting solar wind disturbance at a time when it has traveled well out into interplanetary space. The shape given the

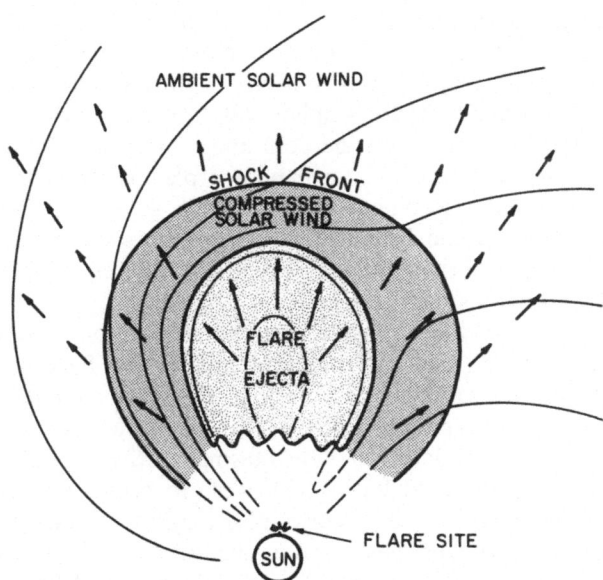

Fig. 6.1   A qualitative sketch, in equatorial cross section, of a flare-produced shock wave propagating into an ambient solar wind similar to that of Fig. 5.9. The arrows indicate the plasma flow velocity and the light lines indicate the magnetic field [6.1]

disturbance in the drawing assumes some lateral expansion; theoretical and observational arguments for this effect will be presented in VI.3 and VI.8. The ambient solar wind plasma and magnetic field lines must be compressed and pushed aside by the expanding flare ejecta (the high electrical conductivity of the plasma prevents rapid interpenetration). If the speed of the ejected material exceeds the ambient solar wind speed by more than the local sound (or Alfvén) speed, a shock front will form at the leading edge of the compressed ambient plasma shell [6.2].

The nature of the boundary between the compressed ambient solar wind and the flare ejecta depends somewhat upon details of the flare process. If the flare region was also a source of the ambient solar wind, the magnetic field lines in the flare plasma must connect to the ambient field. However, for the spatial configuration shown in Fig. 6.1, this connection would take place only on a small part of the boundary, as shown just above and to the right of the flare site. Most of the boundary surface would separate plasmas from different solar source regions. Thus the material on the two sides of the boundary might be expected to have different thermodynamic and chemical properties. Further, magnetic field lines would not cross such a region of the boundary, which must then form a tangential discontinuity within the interplanetary plasma and magnetic field (II.3). If the flare region was not a source of the ambient solar wind, the *entire* boundary surface would separate plasma from different source regions and would be expected to be a tangential discontinuity. (This situation is entirely different from that illustrated by Parker [6.3] for a spherically symmetric wave, where the entire boundary is crossed by the field lines; see also [6.4].) The rarity of collisions in the tenuous interplanetary plasma leads to extremely slow diffusion normal to magnetic field lines (e. g., III.10). The expected tangential nature of the boundary discontinuity would then help to preserve any thermodynamic or chemical differences between the ambient and flare plasmas.

The magnetic field and plasma structure within the material ejected by the flare depends strongly on the details of the flare process. The magnetic field lines must connect back to the flare site (with a current sheet extending through the body of the ejecta as well as along its boundary) unless some diffusion of the plasma relative to the field lines, or "reconnection" of the field lines [6.5], were to occur. Reconnection, a distinct possibility since some theories of solar flares employ this process as the basic flare mechanism, would produce closed magnetic loops within the flare ejecta, as indicated by the dashed field line of Fig. 6.1. This configuration has been advocated by Gold (see the numerous discussions following relevant papers in the collection [6.6]). If some of the ejected material moves outward more rapidly than that near the tangential discontinuity (due either to acceleration of the former or deceleration of the latter) a second shock might form within the flare ejecta. This would be a "reverse" shock, moving toward the sun relative to the plasma but convected outward by the rapid plasma motion [6.4, 6.7].

An example of a solar wind shock has been given in Figs. 2.1 and 2.2 and described in II.3. Such shocks generally occur at the leading edge of a solar wind disturbance involving elevation of the flow speed, proton temperature, and magnetic field intensity, along with a compression (and subsequent rarefaction) in the observed density. These variations are qualitatively consistent with the disturbance structure illustrated in Fig. 6.1. The entire observed post-shock disturbance, which we shall refer to as a "shock wave," persists for 1 to 2 days (II.3). This scale time places the phenomenon in Class 2 or 3 of the classification scheme developed in Chapter II.

In the present chapter we will concentrate on formulating a large-scale description of interplanetary shock waves. Observations of shock wave motions, shapes, plasma flow patterns, and chemical properties will be reviewed using the qualitative expectations of Fig. 6.1 as a guide. One of the major tasks of this review will be to distinguish between flare-produced shocks and those that might form at the fronts of high-speed plasma streams, as mentioned in connection with Figs. 5.10 and 5.11 in V.5. We would hope to isolate the flare-produced class of disturbances for the present discussion. Theoretical models of shock propagation in the solar wind will be described and then used to classify and interpret the observations. Particular attention will be given to the mass and energy contents of interplanetary shock waves, viewed in the context both of the overall energetics of the coronal expansion and of solar flare physical processes.

VI.2 The Motions of Interplanetary Shock Waves Inferred from Plasma Observations

The changes in solar wind properties associated with a shock front occur so quickly as to be "abrupt" when observed by present day plasma detectors. As an example of these changes, Fig. 6.2 shows Vela 4 observations of the solar wind speed, proton density, and proton and electron temperatures (two values of each temperature are given because of the presence of thermal anisotropies). The increases in all four parameters at ~1915 UT signaled the passage of a shock. The observed solar wind properties before and after shock passage are summarized in Table 6.1. The same shock front was observed on the Explorer 34 satellite, from which Ogilvie *et al.* [6.9] have reported similar changes in the proton properties.

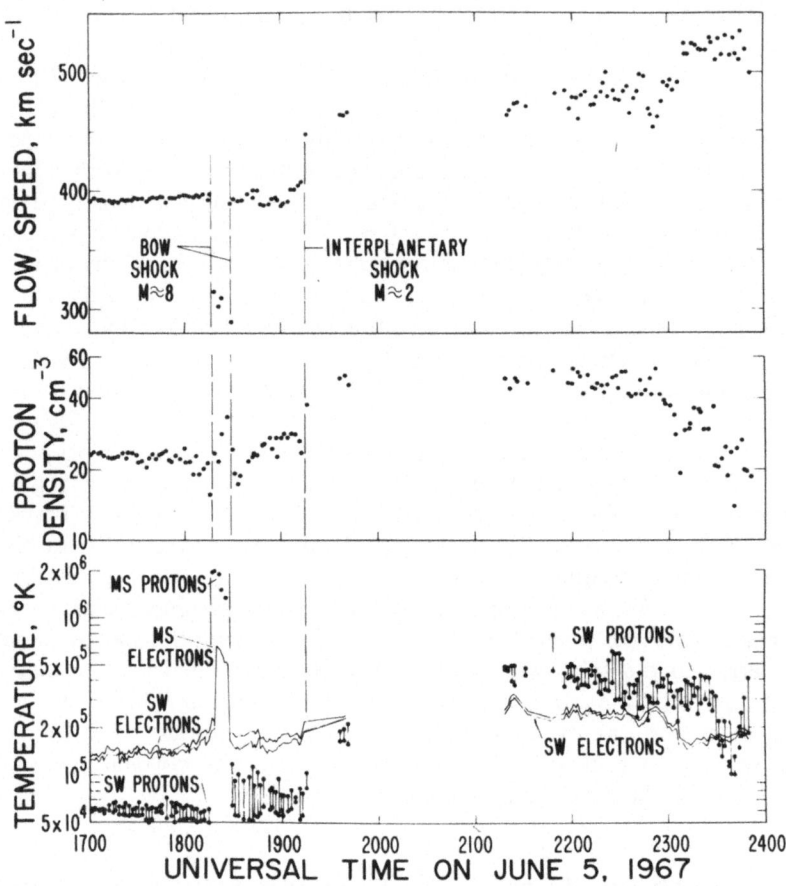

Fig. 6.2　Solar wind properties observed by the Vela 4 B spacecraft on June 5, 1967. The abrupt changes at 1915 UT signal the passage of an interplanetary shock [6.8]

Table 6.1　*Solar Wind Properties Observed near the Shock of June 5, 1967, by a Vela 4 Spacecraft*

	Pre-Shock	Post-Shock
Flow Speed, km sec^{-1}	400	450
Proton Density, cm^{-3}	22	39
Proton Temperature, °K	7×10^4	1.8×10^5
Electron Temperature, °K	1.7×10^5	2×10^5

Much of the interpretation of shock observations, such as that summarized in Table 6.1, is based on the so-called Rankine-Hugoniot relations that express the conservation of mass, momentum, energy, and magnetic flux for the plasma moving through the shock front. If the distribution functions of the important solar wind constituents are assumed to be isotropic and Maxwellian (in a frame of reference moving with the plasma) these conservation laws become [6.10]:

$$\rho_1 v_{1n} = \rho_2 v_{2n}, \tag{6.1}$$

$$\rho_1 v_{1n}^2 + P_1 + \frac{B_1^2}{8\pi} = \rho_2 v_{2n}^2 + P_2 + \frac{B_2^2}{8\pi}, \tag{6.2}$$

$$\rho_1 v_{1n} v_{1t} + \frac{B_n B_{1t}}{4\pi} = \rho_2 v_{2n} v_{2t} + \frac{B_n B_{2t}}{4\pi}, \tag{6.3}$$

$$\left(\frac{1}{2}\rho_1 v_1^2 + \frac{5}{2}P_1 + \frac{B_{1t}^2}{4\pi}\right)v_{1n} - \frac{B_n B_{1t}}{4\pi}v_{1t} = \left(\frac{1}{2}\rho_2 v_2^2 + \frac{5}{2}P_2 + \frac{B_{2t}^2}{4\pi}\right)v_{2n}$$

$$- \frac{B_n B_{2t}}{4\pi}v_{2t}, \tag{6.4}$$

and

$$B_{1n} = B_{2n} = B_n, \tag{6.5}$$

where the flow velocity v is given in the frame of reference moving with the shock front, the subscripts n and t denote vector components normal and parallel to the local shock surface, and the subscripts 1 and 2 denote the pre- and post-shock states. If $B=0$, the resulting hydrodynamic equations permit a single type of shock, for which the normal velocity components v_{1n} and v_{2n} of the inflowing and outflowing material are respectively greater than and less than the sound speeds in the pre- and post-shock gas. In the presence of a magnetic field, the full magneto-hydrodynamic equations permit a complex set of possible shock types. For a summary of these possibilities the reader is directed to a recent review by Burlaga [6.10]. We will be concerned here with the direct analog of the hydrodynamic shock, the so-called "fast shock" in which the normal velocity components v_{1n} and v_{2n} of the inflowing and outflowing material are respectively greater than and less than the "fast-wave" speed $c_f = (c_s^2 + c_A^2)^{\frac{1}{2}}$, where c_s and c_A are the sound and Alfvén speeds, but in which both v_{1n} and v_{2n} are greater than the Alfvén speeds in the pre- and post-shock plasmas. If a sufficient number of plasma and magnetic field parameters are observed at a suspected shock front, equations (6.1) to (6.5) can be used to determine the motion of the shock front (i.e., the frame in which (6.1) holds) and the consistency of the observed thermodynamic and magnetic changes with those expected at a shock. The latter, of course, tests the interpretation of the observed phenomenon as a magnetohydrodynamic shock. Application of this test has usually revealed consistency with the "fast shock" relations to the accuracy (and completeness) of the observations.

As a simple example of the use of Rankine-Hugoniot relations to find the motion of the shock front, consider the observations shown in Fig. 6.2. If the normal to the shock surface is assumed to point radially from the sun, conservation of mass in a stationary frame of reference (neglecting the mass density of all ions other than $^1H^+$) gives

$$n_1(u_1 - U) = n_2(u_2 - U),$$

where U is the propagation speed of the shock relative to a stationary observer. The shock speed is then given by

$$U = \frac{n_2 u_2 - n_1 u_1}{n_2 - n_1}. \qquad (6.6)$$

Use of the Vela 4 (or Explorer 34) observations for the June 5, 1967, shock leads to $U = 510$ km sec^{-1}. Viewed in a frame of reference moving with the shock front, the "upstream" plasma flows into the shock at 110 km sec^{-1}. The sound speed in the pre-shock region is computed to be 58 km sec^{-1}, giving the sonic Mach number of the incoming flow as 1.9.

Several solar flares that occurred a few days before the observation of the June 5 shock at 1 AU could have produced this solar wind disturbance [6.8]. Ogilvie et al. [6.9] have favored association of the shock with a flare observed at 0800 UT on June 3. This would give a transit time from the sun of 59.5 h and a mean propagation speed of 700 km sec^{-1}, appreciably larger than the 510 km sec^{-1} propagation speed inferred from the observations at 1 AU. This difference may indicate a deceleration of the shock in interplanetary space [6.14]. A similar discrepancy could, however, result if the shock normal did not point in the radial direction, as was assumed above.

Table 6.2 summarizes the shock motions deduced either by application of the mass conservation law (as above) or by using the transit time between observation of the same shock front at two separated spacecraft for 27 shocks (presumably all "fast") that have been reported in the literature [6.11]. Although plasma observations are necessary to compute most of the shock parameters of interest here, the observations made with the Imp 3 magnetometer [6.15] are included as they represent the largest available sample of flare associations. The columns in the table give: (1) the date of the shock observation, (2) the sources of observational information, (3) the shock propagation speed, (4) the angle between the radial direction and the observed shock normal, generally determined

Table 6.2 *Properties of Directly Observed Interplanetary Shock Waves*

Date	Source	Shock Speed (km sec⁻¹)	Angle Between Shock Normal and Radial Direction	Sonic Mach Number	Alfvén Mach Number	Transit Time (h)	Mean Speed (km sec⁻¹)
Oct. 7, 1962	Sonett et al. [6.12, 6.13]	510	9°				
Oct. 5, 1965	Gosling et al. [6.14]	420		3.0		17.2 or 71*	2500 or 590*
Oct. 7, 1965	Taylor [6.15]					49.5	840
Jan. 20, 1966	Gosling et al. [6.14]	410		2.3		27.3 or 63.5	1670 or 650
	Taylor [6.15]					63.5	650
March 22, 1966	Chao [6.16]	480	33°		5.5		
March 23, 1966	Chao [6.16]	560	49°		9.2		
July 8, 1966	Taylor [6.15] Lazarus and Binsack [6.17]	750				44.6	930
July 9, 1966	Lazarus and Binsack [6.17]	500					
July 10, 1966	Lazarus and Binsack [6.17]	830					
July 15, 1966	Taylor [6.15]					102	400
Aug. 29, 1966	Taylor [6.15]				4.8	67.2	620
	Chao [6.16]	470	15°				
Sept. 3, 1966	Taylor [6.15]					39.5	1050
Jan. 6, 1967	Taylor [6.15]					55.7	740
Jan. 7, 1967	Taylor [6.15]					45.6	910
Jan. 13, 1967	Bame et al. [6.18]	430*	70°	4*		58	720
Feb. 15, 1967	Hirshberg et al. [6.19]	480*	60°	4.5*		53.5	660

Table 6.2 (continued)

Date	Source	Shock Speed (km sec⁻¹)	Angle Between Shock Normal and Radial Direction	Sonic Mach Number	Alfvén Mach Number	Transit Time (h)	Mean Speed (km sec⁻¹)
May 1, 1967	Hones [6.20]	510*		3*			
May 30, 1967	Ogilvie et al. [6.9]	510				56.2	730
June 5, 1967	Ogilvie et al. [6.9]	510				59.5	700
	Hundhausen [6.8]	350		1.5		?	?
June 25, 1967	Ogilvie and Burlaga [6.21] Lazarus et al. [6.22]	480	8°		3.1		
June 26, 1967	Ogilvie and Burlaga [6.21]	420	33°		8.4		
	Chao [6.16]	500	85°		15		
Aug. 11, 1967	Lazarus et al. [6.22]	340	22°		1.5		
Aug. 29, 1967	Ogilvie and Burlaga [6.21]	420	47°		1.0		
	Chao [6.16]	500	6°				
Sept. 13, 1967	Ogilvie and Burlaga [6.21]	520	27°		1.7		
Sept. 19, 1967	Ogilvie and Burlaga [6.21]						
Jan. 11, 1968	Ogilvie and Burlaga [6.21] Burlaga [6.21]	600	30°		1.3		
Feb. 26, 1969	Bonetti et al. [6.23] Hundhausen et al. [6.24]	570		2.0		?	?

from measurement of the change in magnetic field as will be described in VI.3 (for those observations with no angle given, the shock normal was assumed to be radial in the computation of the propagation speed), (5) the sonic Mach number of the shock, (6) Alfvén Mach number, based on the magnetic field component along the shock normal, (7) the time required for propagation to 1 AU implied by a flare association, and (8) the mean propagation speed corresponding to this time. An asterisk following any entry signifies a value computed by this author from parameters given in the original source or suggested as an alternate value.

The twenty-seven shock observations summarized in Table 6.2 form a reasonable statistical sample from which some average characteristics can be deduced. It should be emphasized, however, that all but one of this sample come from the rising portion of the present solar cycle, and that the characteristics derived therefrom may be typical only of this period. The mean shock propagation speed was $500 \, \text{km sec}^{-1}$, or about $100 \, \text{km sec}^{-1}$ higher than the mean solar wind speed. The shock normals deviated from the radial direction by an average angle of $30°$ (omitting the extreme case of Aug. 11, 1967). The assumption of a radial shock normal in the computation of the shock speed thus leads to an average overestimate of $\sim 15\%$; the value obtained for moderate failures of this assumption is essentially the radial component of the true shock speed. The sonic Mach number was generally near 3, indicating that the typical interplanetary shock was of intermediate strength during this period of observation. The average transit time from the sun implied by flare associations was 55 h; this implies a mean propagation speed inside of 1 AU of $730 \, \text{km sec}^{-1}$. As this latter value is greater than the average propagation speed at 1 AU, a deceleration of the shock during transit from the sun would appear to be common. However, the radial component of the 1 AU shock speed in several cases with large deviations from radial propagation (e. g., Jan. 13 and Feb. 15, 1967) was comparable to the mean speed, indicating that all shocks are not decelerated [6.19] in transit from the sun to 1 AU.

VI.3 Shock Configurations Inferred from Magnetic and Plasma Observations

Some knowledge of the spatial configurations of interplanetary shock fronts is basic to the discussion of shock origins, dynamics, and energetics. For example, we have already seen that shocks may form ahead of either flare ejecta (Fig. 6.1) or long-lived, high-speed plasma streams (Fig. 5.11); it is clear that inclusion of the latter in a discussion of flare-related phenomena could lead to some confusion. The significantly different shock shapes indicated in Figs. 6.1 and 5.11 for the two different types of solar wind disturbances offers a possible means of distinguishing between them. In principle, this knowledge could be easily obtained by observing a single shock wave at several spacecraft spaced at distances comparable to the scale size of the entire disturbance. In practice, few such widely-separated observations have been made [6.1]. Our primary source of information on shock configurations has rather come from statistical analyses of many observations of local shock orientations made at a single position or at closely-spaced positions.

The usual basis for deducing a local shock orientation from observations made at a single spacecraft has been the so-called coplanarity theorem. Application of Maxwell's equations and the momentum conservation laws (6.2) and (6.3) to a compressive shock front in a medium with an isotropic pressure tensor shows [6.4] that the shock normal must lie in the plane defined by B_1 and B_2 (the vector magnetic fields in the pre- and post-shock plasma). Several authors [6.16, 6.25, 6.26] have demonstrated that this theorem remains valid in an anisotropic medium if the pressure tensor is symmetric about the magnetic field (as one would expect on the basis of physical symmetry arguments). As $\Delta B = B_2 - B_1$ must lie in the plane of a shock ($B_{1n} = B_{2n}$), it follows that the shock normal is parallel to $\Delta B \times (B_1 \times B_2)$. Thus, in principle, observation of the pre-shock and post-shock magnetic fields is sufficient to determine a shock orientation. This technique has been applied to actual observations by Sonett et al. [6.13] and Ogilvie and Burlaga [6.21]. Unfortunately, the coplanarity method does not usually lead to an accurate determination of shock orientation; fluctuations in the fields of the pre-shock and post-shock plasmas and the small change in field direction that occurs at many shocks conspire to produce large uncertainties in the computed normal. Ogilvie and Burlaga were forced to use observations from two spacecraft to derive acceptable normals for several shocks despite the availability of magnetic field observations.

The six shock normals derived by Ogilvie and Burlaga [6.21], using both the coplanarity theorem and dual-spacecraft observations, are shown in Fig. 6.3. These normals cluster about the radial from the sun, with a 20° average deviation therefrom. Such a distribution is qualitatively consistent with expectations for the shock configuration of Fig. 6.1, i.e., the flare-produced case. Taylor [6.15] combined the ΔB observed by the Imp 3 magnetometer at eight "possible shocks" (no unambiguous identification was possible due to the lack of plasma data) having reasonable flare associations with the assumption that the normal was parallel to the ecliptic plane. The resulting shock orientations are shown in Fig. 6.4 at a position (on the circle representing 1 AU) corresponding to the solar longitude of each actual observation, relative to the site of the associated flare. All of the shocks except that labeled 101 a are consistent with shock propagation over a broad front roughly symmetric above the flare site. The dashed line on Fig. 6.4 is a circle of radius

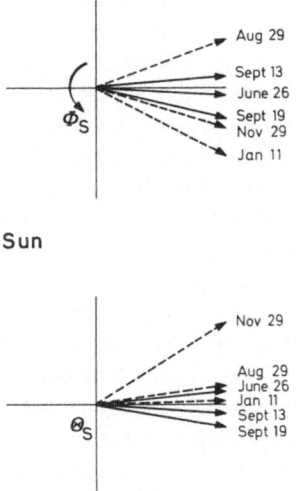

Fig. 6.3   Six interplanetary shock normals derived by Ogilvie and Burlaga [6.21] using the coplanarity theorem (solid arrows) or observations from two spacecraft (dashed arrows). The angle ϕ_s is solar ecliptic longitude, while the angle θ_s is solar ecliptic latitude

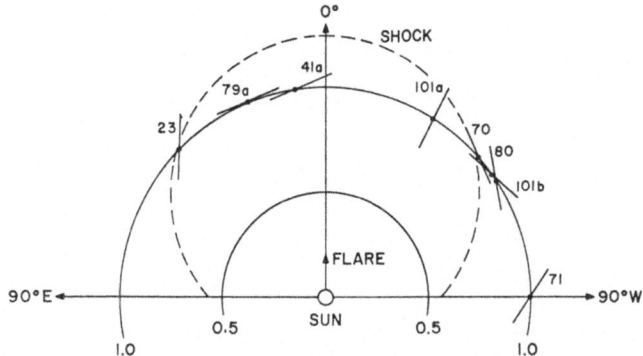

Fig. 6.4   The orientations of eight shock surfaces inferred from Imp 3 magnetometer observations by Taylor [6.15]. Each determination is shown at the longitude of observation relative to the associated flare. The dashed line represents a shock configuration inferred by Hirshberg [6.27] by correlating geomagnetic sudden commencements with solar flares

0.75 AU centered at 0.5 AU, judged by Taylor to be a reasonable representation of the shock configuration implied by the Imp 3 observations.
A similar configuration had been deduced by Hirshberg [6.27] from a
statistical study of geomagnetic sudden commencements and solar flares.

Two more complex techniques for derivation of shock normals have
been proposed to reduce the uncertainties inherent in the coplanarity
method. Chao [6.16] has used the time delay between observations of a
shock at *two different* locations, sufficiently close that the shock front
can be regarded as planar over their separation, to improve the shock
orientations and propagation speeds derived from detailed data obtained at one of the locations. Fig. 6.5 shows five shock normals determined by Chao from Mariner 5 and Pioneer 6 or 7 data. These normals
cluster about an average direction at ~20° from the radial, with a
spread similar to that obtained by Ogilvie and Burlaga (Fig. 6.3). Lepping
and Argentiero [6.28] have combined mass and momentum conservation with Maxwell's equations to derive an overdetermined system of
equations in the plasma densities, flow speeds, and magnetic field components of the pre-shock and post-shock plasmas. A least squares fit of
these equations to plasma and magnetometer data, accumulated before
and after shock passage at a *single* spacecraft, then reduces the effects
of fluctuations (both physical and instrumental) within the accumulation
periods and yields a more accurate shock orientation. This technique
has applied to only a few actual observations.

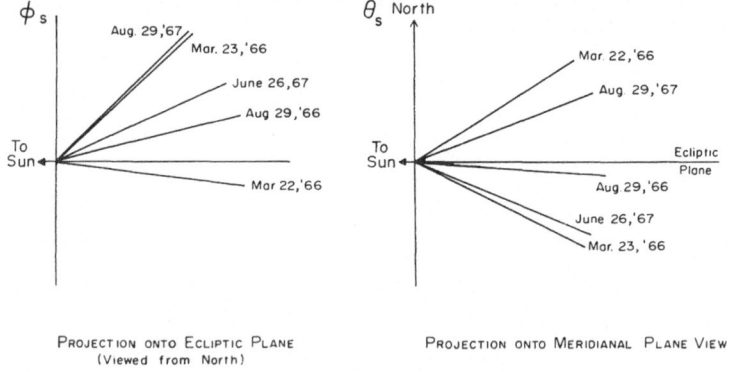

Fig. 6.5   Five interplanetary shock normals derived by Chao [6.16]

Several observers [6.18, 6.19, 6.29, 6.30] have reported examples of
shock normals tilted strongly out of the ecliptic plane, leading to suggestions that the shock fronts may have been more nearly a flattened disc
than a body of revolution about some axis [6.30, 6.31]. This hypothesis
can, of course, be tested by a statistical study of observed shock normals.

The two such studies already described above (Figs. 6.3 and 6.5) lend no support to this idea, showing no greater dispersion of the observed normals out of the ecliptic than in the ecliptic. However, the small number of examples in each study precludes a definitive determination of such details of interplanetary shock configurations. Even our broad conclusion, that the implied configurations more closely resemble those expected from flare-produced disturbances than those expected from long-lived high-speed streams, cannot be stated with great certainty.

VI.4 Characteristics of the Post-Shock Plasma and Magnetic Field

Our qualitative discussion of VI.1 suggested a number of topological features of an interplanetary shock wave; e.g., a shell of compressed ambient solar wind separated from the flare ejecta by a tangential discontinuity, and a reverse shock within the flare ejecta. Further details of the plasma flow within the wave will be predicted by the quantitative theoretical models to be described in VI.8. Most observational attention has been focused on the characteristics of the shock fronts, as discussed in VI.2 and VI.3, and our knowledge of the large-scale topology and flow pattern within the shock wave is limited. Relevant observations that have succeeded in identifying some of the expected structure can be summarized briefly.

Consider first the basic topology of the shock wave. The post-shock compression of plasma, discernible in Figs. 2.1 and 6.2, has generally been found to persist for six to twelve hours. The azimuthal magnetic field component has also been found to be abnormally high over a similar postshock time interval [6.32]. Both of these effects are consistent with the expected structure of the compressed shell of ambient solar wind (Fig. 6.1, or Parker [6.3]). However, the common presence of tangential discontinuities in the interplanetary plasma (II.3) makes identification of the expected interface between this compressed shell and the flare ejecta a difficult task. The best evidence for such a boundary, based on a single example, involves additional information regarding chemical composition, and will be described in VI.5. We thus have only a limited observational indication of the extent of the compressed shell.

Still less information has emerged regarding the structure of the flare ejecta. Hirshberg [6.33] has found that the magnetic fields some twelve or more hours after a shock tend to be intense and variable, with unusually large components normal to the ecliptic plane. There is no evidence concerning the large-scale topology of the field, e.g., the existence of closed magnetic loops. Considerable effort has been

devoted to searches for the "reverse" shock fronts [6.34, 6.35], leading
to the conclusion that such shocks are either rare or difficult to observe.
Of the few examples of reverse shocks that have been found in the solar
wind [6.36, 6.37], only one appears to be related to a flare-associated shock
wave [6.38]. This negative result is rather unexpected, as the general
flow pattern of the post-shock plasma (to be described below) would
appear to favor reverse shock formation. The most commonly observed
feature that may characterize the flare ejecta in a solar wind disturbance
again involves chemical composition and will be discussed in VI.5.

The basic flow pattern of the post-shock plasma has been discussed
more thoroughly. Several studies [6.21, 6.22, 6.35] have identified the
typical post-shock flow as one in which some plasma parameter, either
density, flow speed, or proton temperature, continued to rise after the
abrupt jump observed at shock passage. The examples shown in Figs.
2.1 and 6.2 display this behavior. A more specific classification of interplanetary shock waves has been based [6.35] on the observed temporal
variation of the kinetic energy flux density $f_k = \frac{1}{2}\rho u^3$, where ρ is the
solar wind mass density. Fig. 6.6 shows Vela 3 observations of a Dec. 18,
1965 shock wave (although the abrupt changes at the shock were not
directly observed, the time of shock passage can be inferred with sufficient accuracy for our purposes from the impulsive change in the geomagnetic field observed by ground magnetometers) and the $f_k(t)$ inferred therefrom. The kinetic energy flux density continued to rise above
the immediate post-shock value for $\sim 6\,\mathrm{h}$ after shock passage, only

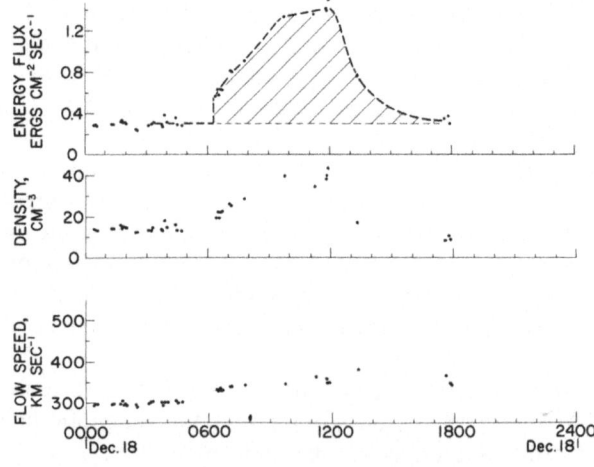

Fig. 6.6   The proton density, flow speed, and kinetic energy flux density observed
before and during the interplanetary shock wave of Dec. 18, 1965 [6.35]. This is the
prototype of the "R-event" in the classification scheme described in the text

thereafter declining to a level comparable to that prevailing before the shock. Interplanetary shock waves of this type, characterized by a continued *rise* in $f_k(t)$ after the abrupt jump at the shock, were designated as "R-events." Fig. 6.7 shows Vela 3 observations of an Oct. 5, 1965 shock [6.14] and the $f_k(t)$ implied by these observations. In this example, the kinetic energy flux density decreased steadily from the immediate post-shock value until the pre-shock level was reached ~18 h after shock passage. Interplanetary shock waves of this type, characterized by a steady *fall* in $f_k(t)$ after the abrupt jump at the shock, were designated as "F-events." In 19 solar wind disturbances classified by this empirical scheme, it was found that the flow speed generally varied slowly, the density more rapidly, in the post-shock plasma. Hence the R or F behavior of $f_k(t)$ was largely determined by the post-shock variations in the density. We will find in VI.9 that empirically defined R and F classes bear some resemblance to two limiting classes of shock waves predicted by theoretical models.

The classification of shock waves described above is largely based on the dynamical behavior in the first 6 to 12 h after shock passage. It thus reflects the structure of what is probably the shell of compressed ambient plasma. The dynamical behavior still later in the shock wave shows less diversity. In both R- and F-events, high-speed, low-density plasma has generally been observed for several days after shock passage [6.35]. The rise in flow speed after ~1200 UT on Fig. 6.7 is an example of this behavior. The energy flux density is not elevated over this time interval due to the prevalance of abnormally low densities. If our con-

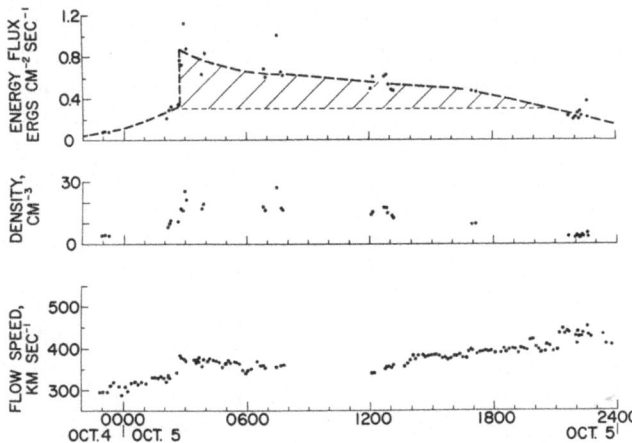

Fig. 6.7   The proton density, flow speed, and kinetic energy flux density observed before and during the interplanetary shock wave of Oct. 5, 1965 [6.35]. This is the prototype of the "F-event" in the classification scheme described in the text

cept of overall shock structure (Fig. 6.1) is at all sound, this pattern of variations would indicate that high-speed plasma is emitted by the corona (but with a "normal" energy flux) for at least several days after a solar flare.

VI.5 The Chemical Composition of the Post-Shock Plasma

One of the most striking features of the post-shock plasma involves the occurrence of "helium abundance enhancements" five to twelve hours after observation of an interplanetary shock or a geomagnetic sudden commencement [6.39, 6.17, 6.18, 6.19, 6.40, 6.41, 6.42, 6.43]. As an example of this phenomenon, Fig. 6.8 shows Vela 3A positive-ion energy-per-charge spectra obtained on Feb. 15—16, 1967. An interplanetary shock wave was detected at ~2345 UT on Feb. 15 both in the Vela 3 plasma data and by the magnetometer on the nearby Explorer 33 satellite [6.19]. Both before and shortly after the arrival of the shock, the observed solar wind helium-hydrogen density ratio was unusually low (0.01 to 0.02), as indicated by the small secondary flux peak in each of the first two spectra of Fig. 6.8. Nine hours after shock arrival the solar wind speed had increased above the immediate post-shock value, but the helium content remained low, as indicated by the third spectrum, obtained at 0841 UT on Feb. 16, of Fig. 6.8. The next solar wind spectrum, obtained at 0916 UT, revealed a drastic change in the secondary flux peak. This spectrum, the fourth in Fig. 6.8, results from a memory mode of data acquisition, with widely-spaced energy channels that preclude a quantitative determination of the helium abundance; however the relative height of the secondary flux peak appears to have increased significantly. A sudden worldwide change in the geomagnetic field occurred on ground magnetometers at nearly this same time, indicating that a *sudden* change in solar wind conditions had taken place. Similar spectra were obtained until 0940 UT when direct data transmission resumed and the fifth spectrum of Fig. 6.8, with normally-spaced energy channels, was acquired. A large secondary peak, closely resembling that in the memory mode spectra obtained since 0961 UT, was present, from which the helium-hydrogen density ratio, h, is computed to be 0.22. The helium content declined in the following twenty minutes to normal values of 0.05. The helium abundance of the solar wind is judged to have been above 0.10 for a total of thirty minutes. The interpretation of this helium enrichment is clarified by the simultaneous observations of the interplanetary magnetic field made on Explorer 33. After the abrupt rise in field magnitude associated with the passage of the shock, unusually high fields were observed until ~0910 UT, when the magnitude abruptly decreased and the field direction shifted. These changes, essentially coincident with the helium enrichment, can be attributed to the passage of a tangential discontinuity in the interplanetary plasma. Hirshberg *et al.* [6.19] have interpreted the simultaneous observation of this discontinuity and the sudden appearance of helium-rich plasma as due to

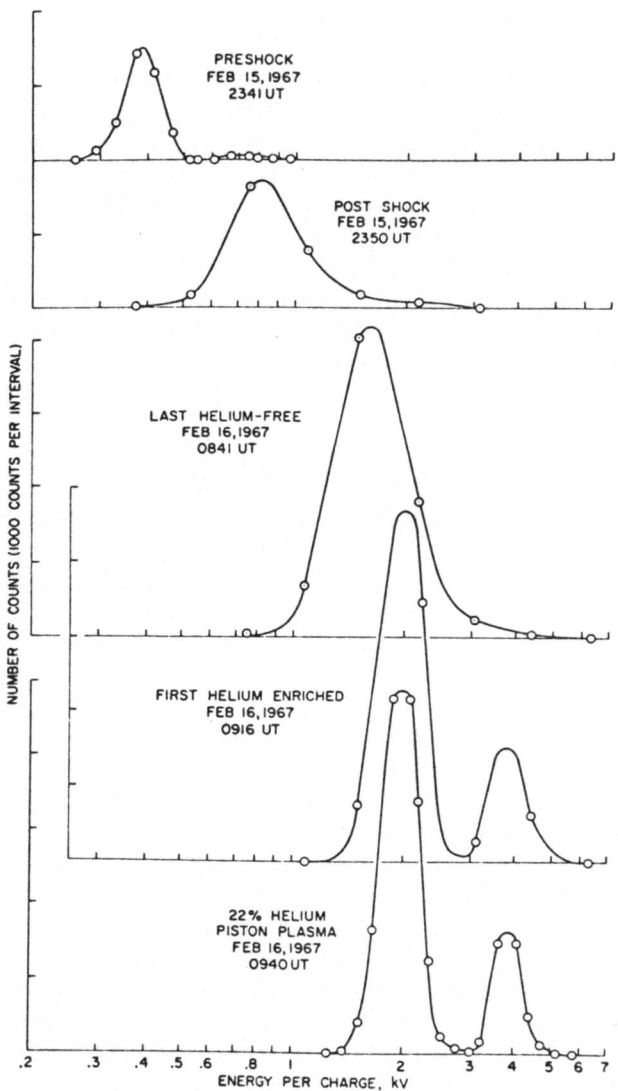

Fig. 6.8   Solar wind energy-per-charge spectra observed by Vela 3A before and during the interplanetary shock wave of Feb. 15—16, 1967 [6.19]

the arrival of the "driver gas" originally ejected by the flare that produced this interplanetary disturbance. The material observed after the shock but before the tangential discontinuity was the compressed ambient solar wind, and thus shared the low helium content of the pre-shock plasma. The flare origin of the material following the tangential discontinuity presumably accounts for its radically different helium content. The region of helium-rich flare ejecta was estimated to be $\sim 10^6$ km thick at the position of observation.

The interplanetary shock and subsequent disturbance of Feb. 15—16, 1967, were probably produced by a very large flare observed optically at ~ 1800 UT on Feb. 13. This and other observations of large helium enrichments after shocks (or sudden commencements) point directly to a flare origin of such events. Hirshberg, Bame, and Robbins [6.42] have examined 16 such helium enhancements ($h \geq 0.15$) observed by Vela 3 spacecraft between July 1965 and July 1967, finding that 12 could be associated with large solar flares. The spatial extent of the enhancements was estimated to be 0.01 to 0.1 AU. This relationship of high inter-planetary helium contents to solar flares and interplanetary shock waves is strong enough to be reflected in 10-day averages of the helium abundance [6.43, 6.11].

These observations have interesting implications regarding the coronal helium abundance. Unless a mechanism can be invoked which accelerates $^4He^{++}$ more efficiently than $^1H^+$, the coronal helium abundance of the material ejected into the solar wind must have been greater than or equal to that observed at 1 AU. The observation of extreme helium enrichments in the solar wind thus implies a similar or greater enrichment in some coronal region. Note that values near 20%, which have been observed in the flare-associated helium enrichments, are consistent with those predicted by the model of Geiss et al. [6.44] in the lower corona (Fig. 4.6). The occurrence of a flare near such a coronal region or layer well might eject some of the helium-rich material into interplanetary space. The magnetic fields in solar active regions are known to be complex with considerable small-scale structure. If such irregular fields are present in the coronal material heated by the flare and thus accelerated to produce the interplanetary shock wave, scattering of particles off the field irregularities could strongly couple ions of differing charge-to-mass ratios, pulling all ions out of the corona with equal efficiency. It is of interest to note that the interplanetary magnetic field in the helium-rich plasma region observed on Feb. 16 (described above) showed large fluctuations in both magnitude and direction [6.19]. The observation of flare-associated interplanetary helium enrichments may then be indirect evidence for the coronal helium concentration discussed in IV.4—IV.6 and summarized in Fig. 4.7.

An alternate interpretation should be mentioned. In fact, none of the plasma detectors that have been used to observe the helium enrichments can differentiate between the ions $^4He^{++}$ and $^2H^+$ (i.e., deuterium). It has been suggested that thermonuclear reactions in hot flare regions might produce $^2H^+$ [6.45, 6.46], and that these ions would be transported to 1 AU in the interplanetary shock wave. There is no conclusive evidence in present-day observations to verify or rule out this possibility.

VI.6 The Relationship between Solar Flares and Interplanetary Shock Waves

The basic concept of the production of interplanetary shock waves by ejection of material from solar flares is a legacy from the era before direct observations were made in interplanetary space. The indirect studies of solar-terrestrial relationships possible in that era led to widely differing views (e.g., [6.47] and [6.48]) as to the physical reality of the apparent connection between flares and interplanetary shocks (inferred from geomagnetic disturbances). Even today, with detailed, direct observations of the shock waves and their characteristics available, the association of an observed shock with a particular flare is not always clear. While some observed shock waves can be reasonably attributed to large solar flares, others can be attributed only to small flares, and some have no reasonable flare associations. A few large flares appear to produce no interplanetary shock waves. The relationship between solar activity and these solar wind disturbances is still, in fact, only imperfectly understood.

One possible source of this confusing situation is the complex nature of solar flares themselves. Flares are commonly reported and classified on the basis of their emission in the Hα line. However, Hα emission is but one in a chain of related physical events; only a few percent of the energy released in the entire process is radiated in this line. Although the widespread availability of flare patrol data based on Hα emission makes these observations useful in attempts to relate flares and shock waves, we should not be surprised if this particular indicator of flare importance fails to yield a full understanding of the physical relationship with interplanetary observations. Other indications of flare-related processes should be included in any thorough study of flares and shock waves.

Let us now illustrate the difficulties encountered in attempting to relate flares and observed shock waves with a few examples. Fig. 6.9 summarizes solar and interplanetary observations from a 27-day solar rotation period in late 1965. Daily values of the Zurich sunspot number R_z and the Ottawa index of 2800 MHz solar radio flux, taken from *Solar Geophysical Data* [6.49], are shown in the first frame. Solar flares observed in Hα and listed in the same compilation are shown in the second frame by vertical lines whose lengths denote optical importance (based on the area of the emitting region). Three-hour averages of the solar wind speed observed by Vela 3 spacecraft are shown in the lowest frame; interplanetary shocks discernible in the Vela data are indicated by a vertical bar (indicating the observed change in flow speed) and the letter S along the flow speed *vs.* time curve. The low level of solar activity during this period can be judged from the low sunspot numbers and radio fluxes. Only fourteen solar flares of importance 1 or greater, including only one flare rated at importance 2 by a single station, were reported during these 27 days. The Vela solar wind observations detected three small interplanetary shock waves; other shocks might have gone undetected during gaps in spacecraft telemetry. None of the three observed shock waves appears to be recurrent [6.35] or associated with a high-speed stream of the nature described in Chapter V. Reasonable flare associations can be proposed for the 3 December and 18 December shocks, but these associations must of necessity involve flares of importance 1 (i.e., flares with Hα emission from a rather small area). Even during this time of low solar activity, there is no clear one-to-one relationship between the observed solar activity and interplanetary disturbances.

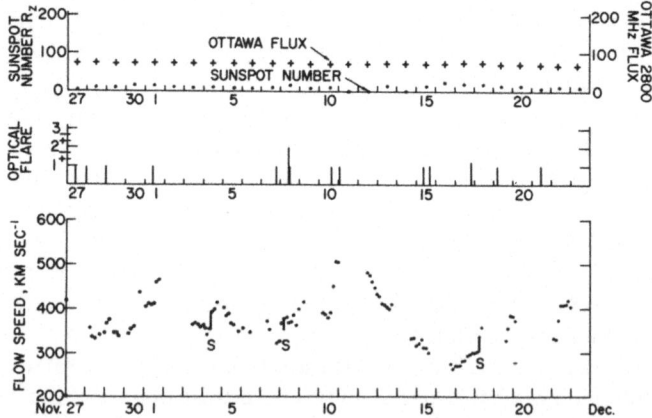

Fig. 6.9 A summary of solar and interplanetary observations made during the 27-day solar rotation period Nov. 27 to Dec. 24, 1965. Zurich sunspot number, Ottawa radio flux, optical flare observations (with the importance index denoted by the length of the vertical line), and three-hour averages of the solar wind speed measured by Vela 3 spacecraft are shown [6.1]

Fig. 6.10 summarizes solar and interplanetary observations from a solar rotation in mid-1967. In addition to the information given in the previous example, Fig. 6.10 also includes type II and type IV radio bursts [6.50] from the compilations in *Solar-Geophysical Data* [6.49] and the *I. A. U. Quarterly Bulletin on Solar Activity* [6.51]. The bursts are shown by vertical lines, dashed for type II, solid for type IV, whose lengths indicate importance on the scale (based on maximum intensity) used in the above sources. Solar activity was at a much higher level during this rotation than during the previous example, as attested by the higher sunspot numbers and 2800 MHz radio fluxes. This difference is manifested in the reporting of 147 importance 1 flares, 12 importance 2 flares, and 2 importance 3 flares during the May-June 1967 solar rotation. It may then be somewhat surprising to find that Vela 3 and Vela 4 satellites detected only four interplanetary shock waves during this rotation. None of the solar wind disturbances related to these shocks appears to be recurrent. Any attempt at flare associations encounters

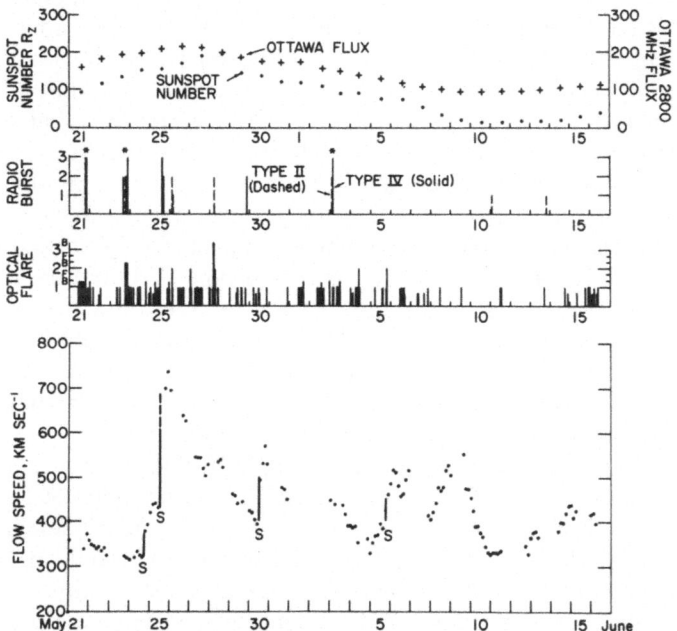

Fig. 6.10 A summary of solar and interplanetary observations made during the 27-day solar rotation period May 21 to June 17, 1967. Type II and type IV radio bursts, indicated by dashed and solid lines (whose lengths denote an importance rating) respectively, have been added to the data displayed in Fig. 6.9. Simultaneous type II and type IV bursts are emphasized by asterisks above the events [6.1]

a problem completely different from the paucity of flares in the previous example; in the present example there are many more flares (even many more large flares) than observed interplanetary shock waves.

Consideration of the radio burst data might be expected to help in clarifying flare associations, as type II and type IV bursts are generally attributed to flare-related coronal processes and have been statistically related to geomagnetic storms. In particular, the occurrence of "a combined type II-type IV burst, which indicates a shock front moving ahead of a plasma cloud through the solar corona" [6.50], has a very high correlation with geomagnetic activity. Three such combinations, hereafter referred to as II-IV radio burst pairs, occurred during the solar rotation under discussion and are indicated on Fig. 6.10 by asterisks above the vertical lines denoting the bursts. Each burst pair can be associated with a solar flare and was followed within three days by an interplanetary shock wave observation at 1 AU.

Use of the radio burst data leads to an entirely reasonable set of flare-radio burst-interplanetary shock associations: an importance 2 N flare and a II—IV burst pair on 21 May with the interplanetary shock observed on 24 May, one of several importance 2 B flares and a II—IV burst pair on 23 May with the interplanetary shock observed on 25 May, and an importance 1 flare and a II—IV burst pair on 3 June with the interplanetary shock observed on 5 June. It is significant that the last of these associations favors an importance 1 flare over two later importance 2 flares as the origin of the June 5 shock. The only remaining interplanetary shock from this rotation period, that of 30 May, can be assigned a reasonable association with an importance 3 flare and simultaneous type II burst (but with no reported type IV burst) on 28 May. Thus use of a combination of optical flare and radio burst data brings some order out of the original chaos, leading to a highly plausible association for each observed interplanetary shock wave.

The relationship between flare-associated radio emission and interplanetary shock waves, suggested by the example given above, appears to have some statistical validity. For example, during the first six months of 1967, seventeen II—IV burst pairs were reported. Nine of the burst pairs were followed within one to three days by an interplanetary shock wave discernible in Vela 3 data; eight of these bursts could be related to simultaneous solar flares that occurred at solar longitudes within 51° of central meridian. Eight of the seventeen reported burst pairs were not followed by observed interplanetary shock waves; of these, three could be related to flares at solar longitudes greater than 50° from central meridian. Thus 60% of the II—IV radio burst pairs related to flares within ~50° of central meridian (a restriction similar to that derived in some indirect studies [6.52]) were followed by an observed interplanetary shock wave. Particular active regions, however, can be completely anomalous [6.1]. A test on the necessity of II—IV burst pairs for the occurrence of interplanetary shock waves (the discussion above tests sufficiency) yields a similar result. During the first half of 1967, nine of the fifteen interplanetary shock waves detected by Vela 3 satellites were preceded (within three days) by reported II—IV radio burst pairs (note that daily gaps do exist in radio spectral observations). This latter test works much less well for the solar rotation of Fig. 6.9. In fact, seven shock observations have been reported from the last half of 1965 [6.1, 6.35], while not one II—IV burst pair is reported from these six months. However, the shock waves observed in late 1965 have been found to be an order of magnitude less energetic than those observed in early 1967 [6.35]; if radio emission were similarly less energetic in 1965, bursts might have occurred but fallen below the threshold of observation. The correlation of radio burst and solar wind observations clearly deserves further and more detailed study.

VI.7 A Synthesis of Interplanetary Shock Observations

The observations described in VI.2 to VI.6 point to several conclusions regarding the nature and structure of interplanetary shock waves. The distributions of observed shock normals (Figs. 6.3 and 6.5) are basically consistent with the configuration expected (Fig. 6.1) for a flare-produced disturbance. The ordering in solar longitude of sudden commencements or observed shock fronts relative to the sites of associated flares (Fig. 6.4) leads to this same consistency. None of these pieces of observational evidence is consistent with the expected steady stream-produced shock configuration (Fig. 5.11). It thus appears that most observed interplanetary shock waves have been produced by solar flares.

Fig. 6.11 is an attempt to synthesize these observations, adding some precision to our earlier qualitative description (Fig. 6.1) of a flare-produced disturbance, hereafter considered to be near 1 AU. The shape of the shock at the leading edge of the disturbance is much as previously drawn. The shock has generally been found to be of intermediate strength, propagating through interplanetary space at ~ 500 km sec^{-1}. The region of compressed ambient solar wind behind the shock is 0.1 to 0.2 AU thick. The tangential discontinuity that is expected to separate the compressed ambient plasma from the flare ejecta (probably observed in one shock wave) is sometimes followed by a thin shell (0.01 to 0.1 AU thick) of helium-rich material. A localized stream of high-speed, low-density solar wind usually follows the flare ejecta; in the two to three days required for the shock wave to reach 1 AU this stream is expected to be distorted into a spiral configuration by solar rotation. Reverse shocks within the flare ejecta or high-speed stream appear only rarely.

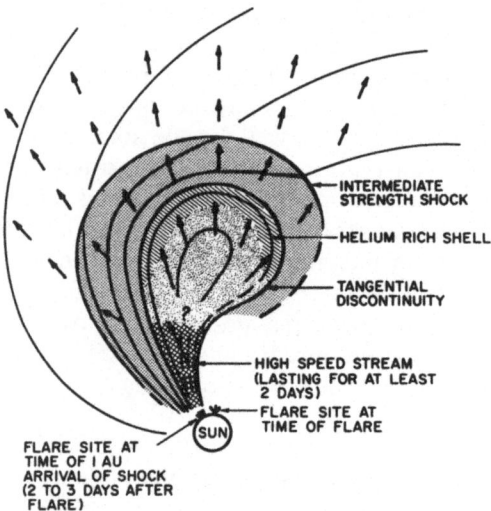

Fig. 6.11 A sketch, in equatorial cross section, of observed features of flare-produced interplanetary shock waves [6.1]

Thus many properties of a flare-produced solar wind disturbance are indicated (although in some cases only tentatively) by presently available observations. Many equally interesting properties remain undetermined. For example, there is virtually no observational evidence related to the possible existence of closed field lines within the flare ejecta (as denoted by the question mark on Fig. 6.11). We emphasize once again that all but one of the observations used in this synthesis date from the rising portion of the present solar cycle. These same observations do indicate some changes within the cycle [6.35]. The large-scale structures of solar wind disturbances might also undergo detailed or even gross changes. Clearly, much remains to be learned from future observations.

VI.8 Theoretical Models of Shock Propagation in the Solar Wind

The theoretical treatment of propagating solar wind disturbances such as shock waves requires integration of the time-dependent equations for mass, momentum, and energy conservation. The presence of the second independent variable, the time, t, so complicates the integrations that numerous simplifying assumptions must be made to yield a tractable problem. All treatments to date have employed a one-fluid formulation, including the forces due to the pressure gradient and solar gravity, and neglecting all energy sources (even heat conduction). Under the further assumptions of a radial, spherically-symmetric flow, the mass, momentum, and energy conservation equations for a gas with $\gamma = 5/3$ become

$$\frac{\partial \rho}{\partial t} + \frac{1}{r^2} \frac{\partial}{\partial r} (\rho u r^2) = 0 , \tag{6.7}$$

$$\rho \left(\frac{\partial u}{\partial t} + u \frac{\partial u}{\partial r} \right) = - \frac{\partial P}{\partial r} - \frac{\rho G M_\odot}{r^2} , \tag{6.8}$$

and

$$\frac{\partial}{\partial t} \left(\frac{P}{\rho^{\frac{5}{3}}} \right) + u \frac{\partial}{\partial r} \left(\frac{P}{\rho^{\frac{5}{3}}} \right) = 0 . \tag{6.9}$$

The last equation simply states that $P/\rho^{5/3}$ is conserved for a given fluid parcel, or that the expansion is adiabatic, as implicitly assumed when all heat sources were neglected. Solutions of this time-dependent problem have been obtained by two different techniques: application of the classical analytic theory of "self-similar" waves [6.53] and direct numerical integration of the equations of motion.

In the first of these techniques, a solution of the system (6.7)—(6.9) is sought in which all fluid properties are a function of the similarity parameter

$$\eta = t r^{-\lambda} \tag{6.10}$$

[6.54, 6.3]; such solutions thus retain the same profile, or dependence on r, at all times. Any fluid parcel moves according to the law

$$r = r_0 \left(\frac{t}{t_0} \right)^{\frac{1}{\lambda}} , \tag{6.11}$$

where r_0 is the position of the parcel at time t_0. The original application of this technique to the problem of shock wave propagation in the solar wind was made by Parker [6.3]. If the gravitational terms were neglected, solutions of (6.7) to (6.9) with a shock front at the leading edge of the wave could be found by numerical quadrature; these solutions required the additional assumption that the flow speed and internal energy of ambient solar wind were negligible (i.e., the shock at the leading edge of the disturbance was strong). In the case where the ambient density varies as $1/r^2$, a good approximation to the interplanetary region, these solutions correspond to shock waves in which the total energy varies as the $3/\lambda - 2$ power of time. Fig. 6.12 shows the density vs heliocentric position (normalized in units of r_s, the position of the shock) for two limiting cases, $\lambda = 1$ and $\lambda = 3/2$. The solution labeled "driven wave" corresponds to $\lambda = 1$ and represents a shock wave in which the energy increases linearly with time. The density rises monotonically behind the shock, becoming infinite as $r \to 0.84$, the vertical line on Fig. 6.12. Both the shock front and the singularity move with a constant outward speed (see 6.11). This wave can thus be thought of as pushed (or "driven") ahead of a steadily expanding "piston" (located at the singularity that does work on the wave at a constant rate). The solution labeled "blast wave" corresponds to $\lambda = 3/2$ and represents a shock wave in which the energy is constant. The density declines monotonically behind the shock. The shock front moves with a steadily decreasing outward speed (proportional to $r^{-1/2}$ from 6.11). This wave can be thought of as produced by a point explosion at $t = 0$, with no further release of energy thereafter. This latter class of solutions (approached by disturbances with the initial energy input persisting for a short time) has an interesting and useful characteristic; the properties of the wave (e.g., the shock speed at or transit time to a given location) depend only upon the total energy of the disturbance.

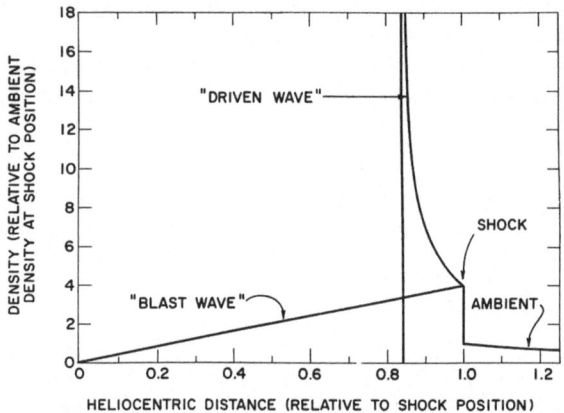

Fig. 6.12 Similarity solutions (density vs heliocentric distance) for interplanetary shock waves. The "driven wave" has an energy increasing linearly with time, while the "blast wave" has a constant energy [6.1]

Extensions of Parker's basic similarity solutions have been carried out by Simon and Axford [6.55], Lee and Balwanz [6.56], Lee and Chen [6.57], and Lee, Chen, and Balwanz [6.58]. Korobeinikov [6.59] has derived similarity solutions in which the assumption of infinite shock strength is somewhat relaxed, his solutions being valid to first order in the ratios of pre- and post-shock flow speeds and temperatures. Dryer [6.60, 6.61] has discussed solutions in the $\lambda = 1$ limit for an infinite strength shock in an ambient medium with finite expansion speed. The latter treatment also includes a finite electrical resistivity, the physical effect of which is to remove the mathematical singularity (in density) at the "piston."

The second technique for studying the propagation of interplanetary shock waves, that of direct numerical integration, can be applied to equations (6.7) to (6.9) with no further simplifying assumptions. Fig. 6.13 shows the density $vs.$ heliocentric position for two disturbances introduced at $r_1 = 0.1\,\mathrm{AU}$ into an adiabatic, steady, ambient solar wind with a flow speed of $400\,\mathrm{km\,sec^{-1}}$ at 1 AU [6.62]. These solutions apply to finite shock strengths, as indicated by the density jump of less than a factor of four (the strong shock jump in a gas with $\gamma = 5/3$) at the shock front. The example labeled "driven wave" was produced by introducing a persistent change in fluid parameters at $r = r_1$. The density rises monotonically behind the shock until a discontinuity, separating the compressed ambient solar wind from the material introduced in the initial disturbance at $t = 0$, is reached. This interface requires special treatment in the numerical integrations, and its properties are only qualitatively indicated on Fig. 6.13. However, there is no density singularity as found at the "piston" interface in the similarity solution for $\lambda = 1$. The wave moves with nearly constant speed, and has an energy increasing linearly with time. It is thus analogous to the driven wave of similarity theory, and can be thought of as a wave pushed by a con-

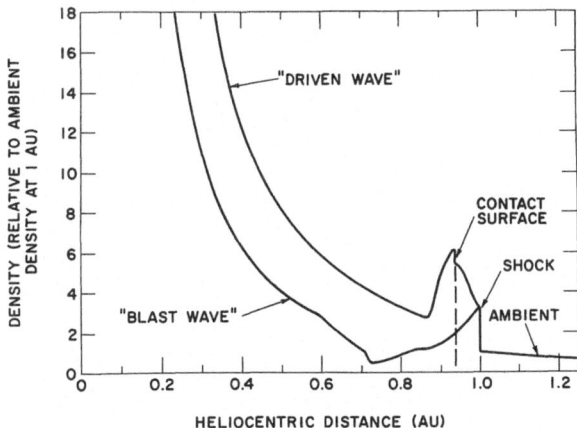

Fig. 6.13   Numerical solutions (density vs heliocentric distance) for interplanetary shock waves. The "driven wave" and "blast wave" cases correspond to the same definitions as in Fig. 6.12 [6.1]

tinuous output of "driver gas" from the sun. A new steady solar wind flow, corresponding to the new boundary conditions at $r=r_1$, has formed behind the compressed shell, as shown for $r \lesssim 0.83$ AU on the figure. The example labeled "blast wave" on Fig. 6.13 was produced by a short duration perturbation in fluid parameters at $r=r_1$. The density decreases monotonically for some distance behind the shock, with an eventual increase to the re-established ambient profile (at $r \approx 0.6$ AU in the figure). This wave moves with a steadily decreasing outward speed and has a constant total (including gravitational) energy. It is thus analogous to the blast wave of similarity theory, and can be thought of as a wave produced by an explosive ejection of material at $t=0$, followed by a return to ambient conditions. As in similarity theory, the properties of this impulsively generated class of disturbances depend only on the total energy [6.62]. The numerical blast waves differ from those of similarity theory in that the density rarefaction does not extend all of the way back to the sun. The numerical solutions involve the release of both mass and energy in the wave, whereas the similarity solutions in the blast wave limit involve only the release of energy.

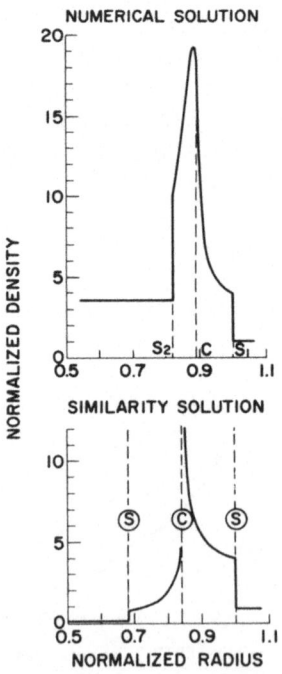

Fig. 6.14 Numerical and similarity solutions showing the existence of a "reverse shock" [6.63]

Fig. 6.14 gives a more detailed comparison of the density (normalized here to the $1/r^2$ ambient density at any heliocentric radius) *vs.* heliocentric position (normalized to one at the leading-edge shock) for the driven wave solutions derived numerically and using similarity theory [6.63]. Both solutions shown involve a new steady flow at small heliocentric distances that is faster than the flow near the contact surface separating the ambient and "driver gas." Both solutions thus include a "reverse shock" (at S_2 in the numerical solution and at the innermost S in the similarity solution) within the driver gas, a possibility mentioned in the qualitative discussion of VI.1. It can be demonstrated using the numerical solutions that this configuration will be observed at a given heliocentric position in interplanetary space only if the initiating solar disturbance persists for more than 10% of the transit time of the resulting interplanetary shock wave to that position [6.63].

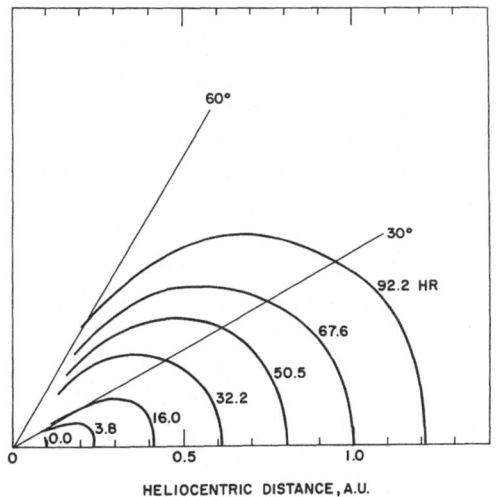

HELIOCENTRIC DISTANCE, A.U.

Fig. 6.15   The shock configuration as a function of time (indicated in hours) produced by a shell of flare ejecta initially confined to a cone of half angle 15° at a heliocentric distance of 0.1 AU [6.64]

The limited spatial extent of optical and radio emission from flares indicates a localized coronal phenomenon. The shock normal observations described in VI.3 imply an interplanetary shock configuration that is not a sun-centered sphere. It would thus appear that, although theoretical models assuming spherical symmetry can approximate the motion of flare-produced shock waves, theoretical models of non-spherical disturbances are necessary to give a detailed description of the shock evolution and spatial configuration. One such model has been obtained for the limited class of flare ejecta confined to a thin spherical shell (at $r=0.1$ AU) within a cone of half-angle θ (with symmetry axis above the flare site) [6.64]. The propagation of this disturbance into a spherically-symmetric, adiabatic ambient solar wind is followed by numerical integration of the hydrodynamic equations analogous to (6.7) through (6.9). As the energy of the disturbance is constant, it is of the "blast wave" class defined above. Fig. 6.15 shows the shape of the resulting shock front at several times (in hours) after introduction of a flare ejecta with $\theta=15°$. The wave slows and expands laterally in propagating to 1 AU; upon arrival at this distance 67.6 h after initiation at 0.1 AU, the shock fills a cone with half angle of nearly 60°. The transverse expansion becomes important when the wave is beyond $r\approx0.4$ AU because interaction with the ambient medium has considerably slowed and weakened the shock; the high pressure produced behind the front can then produce lateral expansion at a significant fraction of the shock propagation speed. An important result of this study is that the shock configuration predicted at 1 AU for disturbances of constant energy is nearly independent of θ for initial half angles in the range $\theta \lesssim 15°$. The limiting shock configuration near 1 AU is a sphere with radius ~0.5 AU, centered at ~0.5 AU along the symmetry axis above the flare site. This shape is in general agreement with that inferred from observations, as displayed on Fig. 6.4.

All of the theoretical models described above (derived by either similarity or numerical techniques) have assumed equal electron and proton temperatures and neglected the thermal conductivity of the electrons. However, Parker [6.65] pointed out that the "thermal equilibration time" (V.5) in the hot plasma behind an interplanetary shock is of the order of 10^4 sec, much shorter than the expected transit time of a flare-produced shock wave to 1 AU. This implies that the flow behind the shock would be more nearly isothermal than adiabatic. If this analysis is extended to more general solar wind conditions, it can be shown that a nearly steady balance will exist between heat conduction and any solar wind heating mechanism persisting on a time scale longer than $\sim4 \times 10^4$ sec [6.66]. In fact, the heat conductivity of interplanetary electrons is so large that only a small electron temperature gradient is required to dissipate the thermal energy released at a typical interplanetary shock. Although the weak nature of the coupling between solar wind protons and electrons (III.5 and III.12) may imply that this "energy sink" has little effect on shock dynamics, heat conduction would be expected to prevent any large rise in the electron temperature in interplanetary shock waves. The observed absence of large temperature variations in observed shock waves (e.g., Fig. 6.1 and Table 6.1) has been attributed to the high electron thermal conductivity [6.66]. Heat conduction would be of great importance in any two-fluid model of shock waves in the solar wind.

VI.9 A Classification of Interplanetary Shock Waves Suggested by the Theoretical Models

The theoretical models described in VI.8 give quantitative descriptions of many shock wave features suggested by our qualitative discussion of VI.1. Some of the general predictions of these models (e.g., the "limiting" configuration for an axially-symmetric shock wave of Fig. 6.15) are in reasonable agreement with observations, while other predictions (e.g., shock speeds at, and transit times to, 1 AU) can be brought into agreement by proper choice of initial conditions. It thus appears that these models give a valid description of the basic dynamical properties of interplanetary shock waves in much the same sense that Parker's early models of a steady, spherically-symmetric coronal expansion (Chapter I) described the basic dynamical properties of the overall coronal expansion. This constitutes an additional piece of evidence for the fluid nature of the interplanetary plasma.

We should not lose sight, however, of the highly idealized nature of the models. None, for example, includes the energy source term due to heat conduction, known to be of fundamental importance (Chapter III) in structureless solar wind models. Even such apparent successes as the prediction of a shock configuration similar to that inferred from observations may be fortuitous. The intermediate strength shocks observed near 1 AU are largely transported through interplanetary space by the general motion of the plasma (at $\sim 400\,\mathrm{km\,sec^{-1}}$) rather than by the motion of the shock relative to the plasma (at only $\sim 100\,\mathrm{km\,sec^{-1}}$). Thus the shock configuration would be grossly distorted from that of Fig. 6.15 by any flow inhomogeneities in the *ambient* solar wind. The detailed fine structure observed in interplanetary shock waves, such as that discernible in Fig. 2.1, may be more related to the general interplanetary abundance of waves and discontinuities than to the structure of the shock wave itself. It would then seem clear that a physically meaningful comparison of theoretical models with observations should concentrate on general features, largely independent of the special assumptions made in formulating the models and the details observed in such solar wind disturbances.

One feasible use of the theoretical models is as a guide to the physical classification of observed shock waves. Both the similarity and numerical solutions, it will be recalled, predict two limiting classes of waves.

The "driven waves" of VI.8, produced by prolonged disturbances near the sun, are characterized by post-shock increases in the density, flow speed, and (in the numerical solutions) temperatures. The "blast waves" of VI.8, produced by impulsive disturbances near the sun, are characterized by post-shock decreases in the density, flow speed, and temperature. For the latter class, both the similarity and numerical techniques give disturbances whose basic properties depend only on the total energy in the wave (despite the fact that the numerical solutions involve a finite mass release while the similarity solutions do not). The numerical solutions predict that the driven wave will be observed at a given position r_I in interplanetary space if the initial disturbance near the sun persisted for more than $\sim 10\%$ of the transit time of the wave to r_I, while the blast wave will be observed at r_I if the initial disturbance persisted for less than $\sim 1\%$ of the transit time [6.62, 6.63]. The two limiting wave forms can thus be related to the duration of the initial disturbance (or flare). The average sun-earth transit time of 55 h for directly observed interplanetary shock waves (VI.2) then implies that interplanetary driven waves should be associated with flares that release energy and mass into the solar wind for a time longer than ~ 5 h, while interplanetary blast waves should be associated with flares that release energy (and perhaps mass) for a time shorter than $\sim \frac{1}{2}$ h. Intermediate types of shock waves would result from flares with intermediate time scales.

The examples of observed interplanetary shock waves shown earlier display post-shock variations in plasma properties that are qualitatively consistent with those in the two classes predicted by the models. The shock waves of Figs. 2.1, 6.2, and 6.6 resemble driven waves, while that of Fig. 6.7 resembles a blast wave. The typical pattern of post-shock variations stated in VI.4, for which at least one of the density, flow speed, or temperature continues to rise after shock passage, suggests that driven or intermediate class waves are most common in the solar wind. The presence of all these classes indicates that interplanetary shock waves are generated by flares that enhance the coronal expansion on time scales ranging from $<\frac{1}{2}$ h to >5 h. The flares with the longer time scales appear to be the most frequent producers of shock waves.

The driven wave and blast wave concepts can also be related to the empirical shock wave classification scheme, based on the post-shock variation of the kinetic energy flux density $f_k(t)$, that was described in VI.4. The theoretical models predict a continued rise in $f_k(t)$ following the abrupt jump at the leading edge of a driven wave, and a monotonic decrease in $f_k(t)$ following the abrupt jump at the leading edge of a blast wave. Thus the driven wave would correspond to the empirical "R-event," and the blast wave would correspond to the empirical "F-event." Although this analogy is tempting and reasonable, all features of observed *F*-events do not fit (even qualitatively) the expected pattern for blast waves. In particular, the observation of elevated flow speeds for several days after many *F*-events (VI.4) is not consistent with the blast wave models. These observations suggest [6.35] a two-stage origin of interplanetary shock waves: a short-lived release of energy followed by a longer emission, from the flare site or some related source, of high-speed, low-density solar wind (with a more normal energy flux). This is a more complex initialization of a shock wave than has been considered in any of the theoretical models.

VI.10 The Mass and Energy in Interplanetary Shock Waves

Two independent techniques have been employed to estimate the energy (and mass) contents of interplanetary shock waves. The first of these uses theoretical models to derive a relationship between the energy in a blast wave and its transit time to or propagation speed at 1 AU. Use of an observed transit time (involving, of course, a flare association) or shock speed then leads to an energy estimate based on the assumption that the shock wave under consideration was of the blast wave class. The second technique involves integration of the observed energy (or mass) flux through a sun-centered sphere with radius 1 AU over the entire duration of an observed shock wave. This latter method has the advantage of being independent of models or flare associations.

For the similarity solution of VI.8 with $\lambda = 3/2$, or the blast wave limit, the fluid parameters in the range of heliocentric distance $0 < r \leq r_s$ (where r_s is the shock position at time t) are given by [6.65]

$$\rho(r) = 4\rho_a(r_s)\frac{r}{r_s},$$ (6.12)

$$u(r) = \frac{3}{4} U \frac{r}{r_s},$$ (6.13)

and

$$P(r) = \frac{3}{4} U^2 \rho_a(r_s) \left(\frac{r}{r_s}\right)^3.$$ (6.14)

U is the propagation speed of the shock front and $\rho_a(r_s)$ is the preshock mass density at $r = r_s$. The total energy in the wave at any time is

$$W = 4\pi \int_0^\infty r^2 \, dr \left(\frac{1}{2}\rho u^2 + \frac{\gamma}{\gamma - 1} P\right).$$ (6.15)

The integrand is zero for $r > r_s$, so that the integration need extend only from 0 to r_s. Substitution of (6.12) to (6.14) into (6.15) gives (for $\gamma = 5/3$)

$$W = \frac{3\pi}{2} \rho_a(r_s) U^2(r_s) r_s^3 \tag{6.16}$$

where r_s, it must be recalled, is a function of time. As the ambient density was assumed to be proportional to r^{-2} and the similarity law (6.11) implies that $U(r_s)$ is proportional to $r^{-1/2}$, W is independent of r_s or of time, as expected for a blast wave. The shock speed at r is then given by

$$U^2 = \frac{2W}{3\pi \rho_a(r) r^3}. \tag{6.17}$$

Integration of (6.20) (recalling $\rho_a(r) \propto r^{-2}$) gives the transit time T of the shock to position r as

$$T^2 = \frac{2\pi \rho_a(r) r^5}{3W}. \tag{6.18}$$

Equations (6.17) and (6.18) are then the shock speed-energy and transit time-energy relationships for blast waves given by the similarity theory. Use of either the mean shock speed of $500\,\mathrm{km\,sec}^{-1}$ or the mean transit time of 55 h inferred from the observations of Table VI.2 leads to an estimate of the mean energy $\langle W \rangle$ in an interplanetary shock wave of 3×10^{32} ergs. Similar values have been deduced from this same basic technique by Korobeinikov [6.59] and by Dryer and Jones [6.67]. The latter have also used a relationship between sonic Mach number and energy (derivable from the strong shock relation 6.16) and a semi-empirical relation between the Mach number and the overpressure $\Delta p/p$ at the shock front (note, however, an error in the coefficient and the exponent of the Mach number in their final result) to deduce somewhat lower values of W from individual shock observations.

The transit time-energy relationship given by the numerical blast wave solutions is displayed in Fig. 6.16. Use of the 55-hour mean transit time leads to an energy of $\langle W \rangle \approx 3 \times 10^{31}$ ergs. The numerically-derived shock speed-energy relationship and the mean shock speed of $500\,\mathrm{km\,sec}^{-1}$ lead to a similar value of $\langle W \rangle$. The order of magnitude difference between the energy estimates based on the similarity and numerical solutions illustrates the model dependence of this technique. The difference is largely due to the neglect of the ambient solar wind speed in the similarity theory used to derive (6.17) and (6.18); a disturbance of a given energy will reach 1 AU sooner if propagating through a rapidly moving ambient solar wind than if propagating through a stationary medium. This effect is important in the solar wind because of the intermediate strength of most observed shocks.

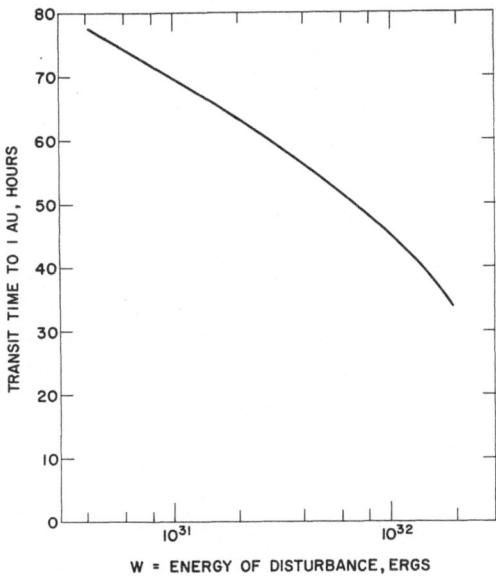

Fig. 6.16 The relationship between energy and transit time to 1 AU for a blast wave propagating in an adiabatic solar wind with $n = 12\,\text{cm}^{-3}$, $u = 400\,\text{km sec}^{-1}$, at 1 AU. The energy for an arbitrary density n at 1 AU is $n/12$ that given by the graph [6.62]

The second technique can be used to estimate either the mass or energy of an interplanetary shock wave from sufficiently complete observations of the mass flux and energy flux densities. Consider, for example, a steady, spherically-symmetric ambient solar wind with mass flux density $\rho_a u_a$ at 1 AU. The mass flux through any sun-centered sphere is $J_a = 4\pi r^2 \rho_a u_a$. Suppose that at time t_0, the mass flux at some inner boundary $r = r_0$ changes to a higher value J', is maintained at this level for a duration $\mathcal{T}$, and then returns to the original value J_a. An extra mass $M = (J' - J_a)\mathcal{T}$ has been added to the flow. The general outward motion of solar wind must eventually transport this perturbation of the ambient conditions to 1 AU, where a variable $J(r_e, t)$ would result. After some time, $J(r_e, t)$ will return to the steady ambient level dictated by the maintenance of J_a at $r = r_0$ for $t > t_0 + \mathcal{T}$. Although the temporal dependence of $J(r_0, t)$ and $J(r_e, t)$ may be quite different, all of the mass added to the flow at $r = r_0$ must eventually pass $r = r_e$. Thus

$$M = \int_0^\infty dt \big(J(r_e, t) - J_a\big) \qquad (6.19)$$

where the integral converges to a finite value because $J(r_e,t) \to J_a$ as $t \to \infty$, as argued above. This result depends only upon the conservation of mass and the continuity of the coronal expansion. Note that a mere redistribution of the material within the spherical shell between $r=r_0$ and r_e would produce a perturbed $J(r_e,t)$ but a *zero* total mass M from the integration in equation (6.19).

A similar argument can be used to demonstrate that the energy in a shock wave passing 1 AU is given by

$$W = \int_0^\infty dt(F - F_a) \tag{6.20}$$

where F is the energy flux (as in equation 3.12) passing through the sun-centered sphere with $r=r_e$. In evaluating the energy in the wave at a different heliocentric position, the change in gravitational potential must be considered. For example, the equivalent energy release at the sun is

$$W_s = W + \frac{GM_\odot M}{r_\odot} \tag{6.21}$$

where M is the mass of the material in the wave.

Equations (6.19) and (6.20) can be written in terms of flux densities as

$$M = \int_0^\infty dt \int dA(\rho u - \rho_a u_a), \tag{6.22}$$

$$W = \int_0^\infty dt \int dA(\tfrac{1}{2}\rho u^3 - \tfrac{1}{2}\rho_a u_a^3) \tag{6.23}$$

where dA is an area element on the sun-centered sphere at $r=r_e$, and where we have neglected all energy flux densities except that due to the kinetic term (a good approximation in the solar wind, as evidenced by Table 3.2). In practice, observations of a particular shock wave are generally available from only one position at $r=r_e$, so that it is necessary to assign an area A to the shock wave and assume that observed values of ρ and u are reasonable averages over this area. Then

$$M \approx A \int_0^\infty dt(\rho u - \rho_a u_a), \tag{6.24}$$

$$W \approx A \int_0^\infty dt(\tfrac{1}{2}\rho u^3 - \tfrac{1}{2}\rho_a u_a^3). \tag{6.25}$$

The remaining integration over time should extend over the entire perturbation from the ambient flux density levels. In actuality, there are no steady ambient levels in the solar wind (see Fig. 5.31), and the integration must be terminated when the mass or energy flux density has approached a nearly constant level close to that observed before the arrival of the shock waves. With these assumptions and approximations, (6.24) and (6.25) can be used to derive purely empirical estimates of the mass and energy in observed interplanetary shock waves.

This technique has been applied to 22 shock waves observed between 1965 and 1967 [6.35, 6.68]. The effective area A was assumed to be $0.7 \times 10^{27} \, cm^2$, or 1/4 of that on the sun-centered sphere. This assumption is consistent with the shock configurations inferred from shock normal observations (Fig. 6.4) and predicted by the model of a non-spherical blast wave (Fig. 6.15). The examples of Figs. 6.6 and 6.7 were included in this study, and the integration over time is indicated in these two figures by the shading under the energy flux density curves (note that these integrations extend into the density rarefaction before the flux densities have returned to the pre-shock levels). The mass estimates thus derived ranged from 5×10^{15} to $1.5 \times 10^{17} \, gm$, with an average $\langle M \rangle = 3.5 \times 10^{16} \, gm$, while the energy estimates ranged from 5×10^{30} to $5 \times 10^{32} \, ergs$, with an average $\langle M \rangle = 7 \times 10^{31} \, ergs$. The equivalent energy release at $r = r_{\odot}$ (equation 6.21) is $\langle W_s \rangle = 1.4 \times 10^{32} \, ergs$. A difference was found between the R- and F-events of the empirical classification scheme (VI.4). For six R-type shock waves, $\langle M \rangle = 3.9 \times 10^{16} \, gm$ and $\langle W \rangle = 6.7 \times 10^{31} \, ergs$, while for six F-type waves, $\langle M \rangle = 1.4 \times 10^{16} \, gm$ and $\langle W \rangle = 1.7 \times 10^{31} \, ergs$. If the R and F classes are identified with driven and blast waves, as suggested in VI.9, the driven waves then transport more mass and energy from the corona than do the blast waves. This might, of course, be due to the longer duration of energy release required to produce driven waves.

Comparison of the purely empirical energy and mass estimates with the model-dependent estimates derived earlier in this section is instructive. The empirical value of $\langle W \rangle = 1.7 \times 10^{31} \, ergs$ for the F-class shock waves, possibly equivalent to blast waves, is an order of magnitude lower than the average energy deduced from similarity theory, and a factor of two lower than the average energy derived from the numerical calculations. The latter of the model-dependent estimates thus appears to be in better agreement with the observations. Note also that a positive mass M is found by integrating the mass flux throughout the F-type shock waves. If these are analogous to blast waves, they involve mass emission, as included in the numerical solutions but not in the similarity solutions. The $M = 0$ shock waves would presumably involve a $J(r_e, t) < J_a$ for a sufficient time interval to compensate for the $J(r_e, t) > J_a$ condition prevailing immediately after shock passage. Within the accuracy of the presently available observations, such a mass flux deficit (not to be confused with a density rarefaction) does not appear to be a common feature of interplanetary shock waves.

VI.11 The Role of Interplanetary Shock Waves in the Overall Mass and Energy Transport from the Corona

The estimates of the mass and energy contents in interplanetary shock waves deduced in VI.10 permit us to assess the role of shock waves in the transport of mass and energy from the solar corona. For this purpose we will use the values $\langle M \rangle = 3.5 \times 10^{16}$ gm and $\langle W_s \rangle = 1.4 \times 10^{32}$ ergs estimated from the integration of observed mass and energy flux densities. If the scale time of emission is taken to be a few hours (say 10^4 sec), as suggested by the comparison of observed post-shock variations with the predictions of theoretical models (VI.9), very high rates of coronal mass and energy losses are associated with flare-produced shock waves; $J = 3.5 \times 10^{12}$ gm sec^{-1} and $F = 1.4 \times 10^{28}$ ergs sec^{-1} respectively. These values should be compared with the loss rates obtained if the "quiet," ambient solar wind summarized in Tables 3.1 and 3.2 were to be emitted from the entire corona—$J_a = 1 \times 10^{12}$ gm sec^{-1} and $F_a = 3 \times 10^{27}$ ergs sec^{-1}. Thus mass and energy are added to the solar wind from the relatively small region thought to be involved in the flare process at rates greater than that from the entire corona under quiet conditions.

An evaluation of the role of flare-produced shock waves in the overall coronal expansion process requires some information as to the frequency of their occurrence. The time intervals summarized in Figs. 6.9 and 6.10 appear to be typical of the rising portion of the present solar cycle, with an average of 3 or 4 shock waves observed per 27-day solar rotation. If these shocks propagate within a cone of $\sim 60°$ half angle, as indicated by Figs. 6.4 and 6.15, roughly 1/3 of all shocks would be observed at a given position in interplanetary space, suggesting that no more than 10 such events occur during a solar rotation (some small shock waves might escape detection, but would presumably carry only small amounts of mass and energy). Thus reasonable upper limits on the mass and energy carried away from the corona by flare-produced shock waves during a typical solar rotation are $M \approx 3 \times 10^{17}$ gm and $W_s \approx 1 \times 10^{33}$ ergs. The mass and energy losses in a spherically-symmetric expansion of "quiet" ambient solar wind (Tables 3.1 and 3.2 once again) for the same 27-day interval would be $M_a \approx 2 \times 10^{18}$ gm and $W_{sa} \approx 1 \times 10^{34}$ ergs. We are thus led to a conclusion similar to that reached in considering the mass and energy losses in high-speed plasma streams (V.8). Although the mass and energy transport in flare-produced shock waves is by no means negligible, it does not appear to dominate the overall efflux of mass or energy in the coronal expansion. Montgomery et al. [6.68] arrived at this same conclusion from detailed examination of solar wind observations made between 1965 and 1967; shock waves generally accounted for $\sim 10\%$ of the total mass and energy transport. This result is in accord with the lack of any apparent relation between long-term trends in solar activity and observed solar wind mass and energy flux densities, displayed in Fig. 5.34. Despite the high rates of emission possible in flare-related disturbances, their relative rarity has limited them to a minor role in the overall coronal expansion during that time interval for which we have detailed solar wind observations.

VI.12 Implications Regarding the Physics of Solar Flares

The plasma emitted by a solar flare, observed in interplanetary space as part of a shock wave, carries information regarding flare processes. Thus, in principle, it can be added to the emissions traditionally considered (e. g., optical and ultra-violet radiation, X-rays, radio bursts, and energetic particles) in the observational study of these processes. In practice, the interaction of the ejected plasma with the ambient solar wind complicates the interpretation of interplanetary shock observations, and requires careful combination of the observations with theoretical models in any such study. Much of the discussion of this chapter has concerned attempts to forge such a combination, and it should be clear that no complete, definitive interpretation of shock wave observations is, as yet, possible. Nonetheless, the estimates of the masses and energies released into the solar wind by flares and the time scale of this process have some interesting implications regarding flare phenomena. In pursuing these implications, we should bear in mind the possibility that our conclusions may not apply to flares in general. In considering flares that produce interplanetary shock waves we may have selected a special class of events.

Of the 22 interplanetary shock waves whose mass and energy contents have been estimated [6.35, 6.68], 8 have plausible associations with optical flares accompanied by type II and/or type IV radio bursts. Table 6.3 summarizes this set of solar-interplanetary associations. Six of the flares were of importance 2 or greater, and could thus be described as large flares. The average duration of $H\alpha$ emission from the eight flares was 115 min, comparable to the several-hour time scale for energy deposition deduced from interplanetary shock profiles in VI.9 (as well as typical of large flares [6.69]). The average mass of the associated solar wind disturbances was 5.6×10^{16} gm, while the average equivalent energy at 1 solar radius was 2.3×10^{32} ergs.

The $H\alpha$ emission from a large solar flare comes from a volume of $\sim 10^{28}$ cm^3, wherein the electron density is $\sim 10^{13}$ cm^{-3} [6.69]. Thus the mass within the luminous volume is $\sim 2 \times 10^{17}$ gm. Bruzek [66.9] has estimated the mass in visible flare-associated ejections to be $\sim 2 \times 10^{16}$ gm. Thus the mass ejected into interplanetary space, $\sim 5 \times 10^{16}$ gm as

Table 6.3   *Associations of Flares and Radio Bursts with Interplanetary Shock Waves of Known Mass and Energy*

Date (1967)	Flare			Radio Bursts		Interplanetary Shock wave			Energy at Sun (erg)
	Imp.	Duration	Position	Type	Duration	Date	Time	Mass (gm)	
Feb. 4	2	1641—1902	N11E40	II IV	1708—1728 1705—1846	Feb. 7	1640	4×10^6	1.6×10^{32}
Feb. 13	3	1749—2130	N20W10	II IV	1803—1820 1829—2438	Feb. 15	2345	5×10^{16}	1.9×10^{22}
May 3	2B	1537—1926	N25E51	II IV	1548—1603 1603—1650	May 7	0100	3×10^{16}	0.9×10^{32}
May 21	2N	1919—1945	N24E39	II IV	1923—1945 1923—2100	May 24	1730	4×10^{16}	1.1×10^{32}
May 23	(3 Possible 2B Flares)		N30E25	II IV	1846—1905 1843—1900	May 25		5×10^{16}	5.0×10^{32}
May 28	3B	0527—0712	N28W33	II	0545—0552	May 30	1430	7×10^{16}	3.3×10^{32}
June 3	1N	0243—0342	N24E14	II IV	0243—0250 0235—0450	June 5	1915	4×10^{16}	1.1×10^{23}
June 23	1N	0039—0108	N15E34	II IV	0053—0110 0110—0203	June 25		1.3×10^{17}	3.7×10^{32}

deduced above, is roughly equal to that in the visible ejections and an appreciable fraction ($\sim 1/4$) of that within the flare region (note that these conclusions differ from those of Bruzek, who estimated a much smaller mass in the interplanetary shock wave). Mass ejection on this scale must be expected to produce large changes in the chromosphere and corona near the flare site. Some evidence for shock-related changes in the coronal density structure has been found from optical observations [6.70]. The *addition* of this large amount of mass to the normal wind flow suggests (but does not prove) the ejection of material from a region not previously participating in the coronal expansion, such as the closed field lines above a complex active region. The coronal magnetic configuration computed above active regions using a source surface model (V.6) has, in fact, been found by Valdez and Altschuler [6.71] to change from closed to open following some large flares. This result implies a flare-associated change so drastic as to modify the *photospheric* magnetic fields upon which the source surface model is based.

Table 6.4   *Estimated Energy Releases in a Large Solar Flare*

Process	Energy (ergs)
$H\alpha$ Emission	10^{31}
Line Emission (including $H\alpha$)	5×10^{31}
Continuum Emission	8×10^{31}
Total Optical Emission	10^{32}
Soft X-rays (1 to 20 A)	2×10^{30}
Radio Bursts	10^{25}
Energetic Protons ($E > 10\,\mathrm{MeV}$)	2×10^{31}
Solar Cosmic Rays ($E = 1$ to 30 GeV)	3×10^{30}
Visible Ejections	10^{31}
Interplanetary Shock Wave	2.3×10^{32}

Table 6.4 compares the average energy at 1 solar radius in an interplanetary shock wave with other energy losses from a large flare; the tabulation is based on the energy estimates (some valid only to an order of magnitude) given by Bruzek [6.69], with correction of typographic errors and adoption of our shock wave energy estimate. All loss processes other than optical emission and the interplanetary shock wave are negligible despite their interest as indicators of physical phenomena. The shock wave appears to carry away at least half of the energy released in a flare, with most of the remainder radiated in the visible wavelengths. This is in contrast to the normal energy balance of the chromosphere and corona (III.7), wherein only about 1% of the energy loss is due to the

ambient solar wind flow. Recent estimates of the continuum emission for the white light flare of May 23, 1967 (included among the flare associations of Table 6.3) give an energy loss of only 6×10^{29} ergs [6.72], suggesting that the value in Table 6.4 is an overestimate. If this is the case, the shock wave appears to carry away a still larger fraction of the energy released by the flare. The total energy release must be at least 2×10^{32} ergs. This, in turn, requires a specific energy release of $\sim 2 \times 10^4$ ergs cm^{-3} from the luminous volume of a flare. A comparison with the normal chromospheric thermal energy density of 5 ergs cm^{-3} and the normal coronal energy density of 1 erg cm^{-3} illustrates the magnitude of the problem to be faced in any model of energy storage and release by a solar flare.

Chapter VII

Concluding Remarks

VII.1 Neglected Topics

Our decision to concentrate in this volume on large-scale solar wind
phenomena, with emphasis upon the relationship of these phenomena
to solar structures and processes, has led to the neglect of several interesting and important areas of solar wind research. Some mention
of such areas, along with recent reviews and key works that will serve
to guide the reader interested in their exploration, is in order. Introduc-
tory treatments of many of these topics have been given by Brandt [7.1].

(1) Techniques of Solar Wind Observations

General discussions of the techniques by which *in situ* observations
of the interplanetary plasma and magnetic field are performed have been
reviewed by Bernstein [7.2], Hundhausen [7.3], Vasyliunas [7.4], and
Ness [7.5]. General surveys of indirect methods have been presented by
Lust [7.6] and Axford [7.7]. For more specific discussions of such
methods see Brandt [7.1] on comet tail observations, Vitkevich [7.8]
and Hewish [7.9] on radio scintillations, and James [7.10] on radar
studies of coronal motions.

(2) Small-Scale Interplanetary Phenomena

The basic theory and observations of hydromagnetic discontinuities in
the solar wind have been reviewed by Burlaga [7.11]. The relative importance of tangential and rotational discontinuities in the interplanetary
plasma (II.3) is considered in recent papers by Burlaga [7.12], Turner
and Siscoe [7.13], and Belcher and Davis [7.14]. Plasma waves and
"turbulence" are discussed in Scarf [7.15].

(3) Extensions of the Coronal Expansion Concept

The generalization of the basic hydrodynamic theory of the coronal
expansion to stars other than the sun is discussed by Parker [7.16] and

by Holzer and Axford [7.17]. Banks and Holzer [7.18] have also described the flow of material from the earth's polar ionosphere in terms of a fluid expansion in some ways resembling that of the corona.

(4) Solar Wind Interactions with Other Particles, Fields, and Planetary Bodies

One aspect of the interaction of the solar wind with the interstellar gas and magnetic field was mentioned in IV.9. For a thorough discussion of this topic (including the "termination" of supersonic flow) see Axford [7.19]. The effects of the interplanetary magnetic field on cosmic rays and solar energetic particles (with use of the latter to probe the field line topology) has been extensively studied; see Axford [7.19, 7.20], and Jokipii [7.21]. The interactions of the solar wind with the moon and planets have been reviewed by Ness [7.22] and by Spreiter and Alksne [7.23]. The solar wind-geomagnetic interaction is perhaps the most complex (and least understood) phenomenon that we have mentioned. For a whole volume of recent reviews, see Dyer and Roederer [7.24].

Some work related to the large-scale structure of the solar wind but only loosely connected to the particular topics selected for discussion in Chapters III to VI should be at least briefly mentioned.

(5) Formal Aspects of the Solar Wind Equations

The formal nature of the solutions to the fluid equations for a steady, spherically-symmetric coronal expansion remains the subject of extensive analysis. A few recent developments concerning the possible types of "one-fluid" solutions were noted in III.4. Other recent work on this topic has been published by Weber [7.25], Dahlberg [7.26], Yeh [7.27], Durney and Roberts [7.28], Durney and Werner [7.29], and Roberts [7.30]. The solar wind properties predicted by the one-fluid solutions with $T \propto r^{-4/3}$ at large r are compared with observations by Durney [7.31].

(6) Heliographic Latitude Dependencies of Solar Wind Properties

Our discussion of a coronal expansion that deviates from spherical symmetry was limited to consideration of long-lived, localized, high-speed plasma streams (Chapter V). Recent analyses of interplanetary magnetic field observations have suggested a heliographic latitude dependence in the sign of the radial [7.32] and out-of-the-ecliptic field components [7.33]. Delineation of any such effect is inherently difficult because present space probe observations are limited to the ecliptic plane and thus to the range $\pm 7°$ of heliocentric latitude. Rosenberg [7.34] has interpreted the proposed latitude dependencies as an interplanetary extension of a solar dipole field. If the reality of this latitude structure is borne out (it is not universally accepted [7.35] at present), another large scale solar wind structure will invite discussion. A relevant "three-dimensional" model of the coronal expansion is considered by Suess [7.36].

VII.2 Solar Activity and the Solar Wind

In Chapters V and VI we discussed at some length two important classes of large-scale solar wind phenomena; long-lived, high-speed plasma streams and flare-produced, transient shock waves. Both of these phenomena are often thought of as "disturbances" superposed on an otherwise structureless coronal expansion. This view (partly motivated by the historical considerations briefly described in II.3) suggests interpretation of these solar wind disturbances as the interplanetary manifestations of solar activity (and thus the first step in the chain of physical associations that comprise the field of solar-terrestrial relations). Our synthesis of observations and theory led to tentative conclusions regarding the relationship between solar activity and these most conspicuous of interplanetary disturbances. The importance and somewhat unexpected nature of these conclusions justifies their reiteration as a closing statement regarding our present understanding of the relationship between solar activity and the solar wind.

The identification of the solar sources for the high-speed plasma streams observed at 1 AU (and by implication of the long-sought "M-regions" responsible for recurrent geomagnetic activity) has remained a difficult task. However, a reasonable case can be drawn (V.6 and V.9) for the contention that these streams arise from large regions of unipolar, diverging coronal magnetic fields. These magnetic regions need bear no direct relationship to solar active regions. The coronal structure and plasma flow implied by this conclusion are sketched in Fig. 7.1. This physical picture is consistent with the "cone-of-avoidance" school of thought regarding the nature of M-regions. In fact, Fig. 7.1 differs only in details from the sketch of Billings and Roberts displayed as Fig. 5.33. It lends little support to the existence of a physical relationship between the plasma streams and centers of solar activity.

The association of solar flares with directly observed interplanetary shock waves has generally been possible, and the physical relationship between these two phenomena remains widely accepted. However, it may be surprising to have found little or no evidence for a large change in the rate at which interplanetary shock waves were observed during the rising portion of the present solar cycle. It is also clear that all large (as observed in Hα) flares do not produce interplanetary shock waves. These findings, along with the observational inference of very large mass *additions* to the solar wind flow in association with shock waves, suggest (1) that flares occur within magnetic structures that are normally "closed" and not related to the solar wind flow, and (2) that only occasionally is the flare process so energetic as to overcome magnetic forces and "open" the magnetic structure. Fig. 7.2 is a sketch of the transient modification of the coronal structure in Fig. 7.1 that might be expected when the flare ejecta and shock wave do escape into interplanetary space. The normal observation of high-speed solar wind for several days after passage of a shock may indicate that this modification persists on such a time scale.

The conclusions summarized above (which should be clearly labeled as speculative) tend to deemphasize any common and strong connection between solar activity and the solar wind. This implication is entirely consistent with a relevant observation—that the parameters characterizing the solar wind flow, mass transport, and energy transport have not displayed any marked, long-term trends during the present solar cycle (Fig. 5.34). In fact, the "solar wind disturbances" that we have discussed appear to play only a small role in the overall transport of mass and energy by the solar wind (V.8 and VI.11). We are thus led to conclude that the basic coronal expansion is not an effect of solar activity, and may even occur from those coronal regions not directly

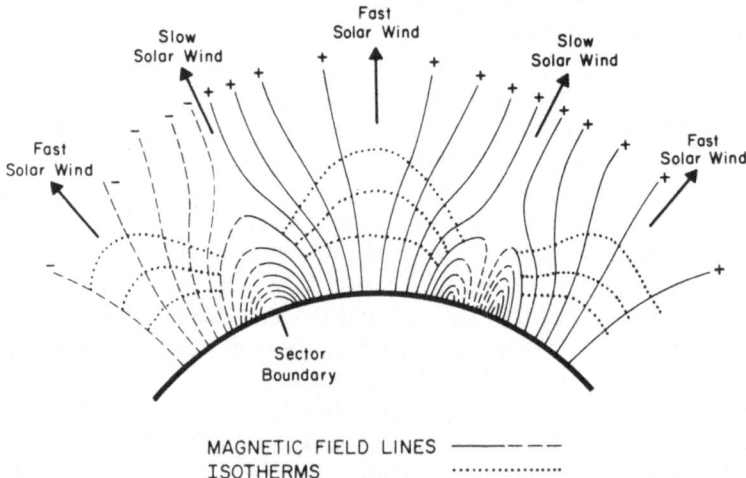

MAGNETIC FIELD LINES ———— — —
ISOTHERMS   ····················

Fig. 7.1   A qualitative (and speculative) sketch of the coronal structure responsible for high-speed plasma streams. This picture is similar to that of Billings and Roberts (Fig. 5.33) except in details (such as the extension of isotherms in closed field regions)

related to activity. Our conclusions do argue for a strong connection
between the solar magnetic field and the solar wind; solar flares are
generally regarded as the release of energy in the solar magnetic field,

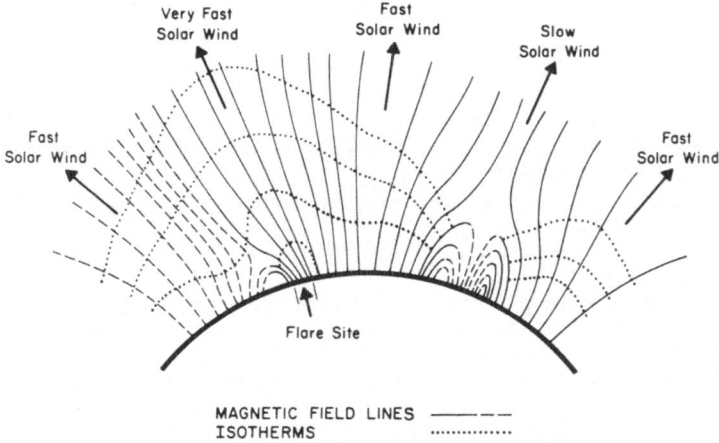

Fig. 7.2   A qualitative sketch of the transient modification, produced by a flare, of
the coronal structure shown in Fig. 7.1. It has been assumed that the flare was
sufficiently energetic to overcome magnetic forces and "open" a region of previously
closed field lines

and the channeling of plasma flow by the coronal field for both shock
waves and steady, high-speed streams has been deduced above. If our
deliberations have led to any single point of emphasis, that point is the
importance of the interaction between the coronal plasma and magnetic
field in the formation of large-scale solar wind structure.

References

Chapter I

1.1. Chapman, S.: An outline of a theory of magnetic storms. Proc. Roy. Soc. (London) Ser. A **95**, 61–83 (1919).
1.2. — Ferraro, V.c.A.: A new theory of magnetic storms. Terr. Magn. **36**, 77–97 (1931).
1.3. Dessler, A.J.: Solar wind and interplanetary magnetic field. Rev. Geophys. **5**, 1–41 (1967).
1.4. Behr, A., Siedentopf, H.: Untersuchungen über Zodiakallicht und Gegenschein nach lichtelektrischen Messungen auf dem Jungfraujoch. Z. Astrophys. **32**, 19–50 (1953).
1.5. Biermann, L.: Kometenschweife und solare Korpuskularstrahlung. Z. Astrophys. **29**, 274–286 (1951).
1.6. — Physical processes in comet tails and their relation to solar activity. Extrait des Mem. Soc. Roy. Sci. Liege Collection in −4° **13**, 291–302 (1953).
1.7. Spitzer, L.: Physics of Fully Ionized Gases. New York: Interscience Publishers 1956.
1.8. Chapman, S.: Notes on the solar corona and the terrestrial ionosphere. Smithsonian Contrib. Astrophys. **2**, 1–11 (1957).
1.9. Parker, E.N.: Dynamics of the interplanetary gas and magnetic fields. Astrophys. J. **128**, 664–675 (1958).
1.10. — Interplanetary Dynamical Processes. New York: Interscience Publishers 1963.
1.11. Chamberlain, J.W.: Interplanetary gas III. A hydrodynamic model of the corona. Astrophys. J. **133**, 675–687 (1961).
1.12. Pannekoek, A.: Ionization in stellar atmospheres. Bull. Astron. Inst. Neth. **1**, 107–118 (1922).
1.13. Rosseland, S.: Electrical state of a star. Monthly Notices Roy. Astron. Soc. **84**, 720–728 (1924).
1.14. Chamberlain, J.W.: Interplanetary gas II. Expansion of a model solar corona. Astrophys. J. **131**, 47–56 (1960).
1.15. Gringauz, K.I., Bezrukikh, V.V., Ozerov, V.D., Rybchinskiy, R.E.: Study of the interplanetary ionized gas, high energy electrons, and solar corpuscular radiation by means of three electrode traps for charged particles on the second Soviet cosmic rocket. Soviet Phys. "Doklady" (English Transl.) **5**, 361–364 (1960).
1.16. — Some results of experiments in interplanetary space by means of charged particle traps on Soviet space probes. Space Res. **2**, 539–553 (1961).
1.17. — Bezrukikh, V.V., Musatov, L.S.: Solar wind observations with the Venus 3 Probe. Cosmic Research **5**, 216–222 (1967).
1.18. Bonetti, A., Bridge, H.S., Lazarus, A.J., Lyon, E.F., Rossi, R., Scherb, F.: Explorer 10 plasma measurements. J. Geophys. Res. **68**, 4017–4063 (1963).
1.19. Scherb, F.: Velocity distributions of the interplanetary plasma detected by Explorer 10. Space Res. **4**, 797–818 (1964).

1.20. Snyder, C.W., Neugebauer, M.: Interplanetary solar wind measurements by Mariner 2. Space Res. **4**, 89–113 (1964).
1.21. Neugebauer, M., Snyder, C.W.: Mariner 2 observations of the solar wind, 1, Average properties. J. Geophys. Res. **71**, 4469–4484 (1966).
1.22. — — Mariner 2 observations of the solar wind, 2, Relation of plasma properties to the magnetic field. J. Geophys. Res. **72**, 1823–1828 (1967).
1.23. Ness, N.F., Scearce, C.S., Seek, J.B.: Initial results of the Imp 1 magnetic field experiment. J. Geophys. Res. **69**, 3531–3569 (1964).

Chapter II

2.1. Parker, E.N.: Interplanetary Dynamical Processes. New York: Interscience Publishers 1963.
2.2. Burlaga, L.F.: Directional discontinuities in the interplanetary magnetic field. Solar Phys. **7**, 57–71 (1969).
2.3. — Ness, N.F.: Macro- and micro-structure of the interplanetary magnetic field. Can. J. Phys. **46**, 5962–5965 (1968).
2.4. Hundhausen, A.J., Bame, S.J., Asbridge, J.R., Sydoriak, S.J.: Solar wind proton properties: Vela 3 observations from July 1965 to June 1967. J. Geophys. Res. **75**, 4643–4657 (1970).
2.5. Roederer, J.G.: Dynamics of Geomagnetically Trapped Radiation. Berlin-Heidelberg-New York: Springer 1970.
2.6. Neugebauer, M., Snyder, C.W.: Mariner 2 observations of the solar wind, 1, Average properties. J. Geophys. Res. **71**, 4469–4483 (1967).
2.7. Coleman, P.J., Jr., Davis, L., Jr., Smith, E.J., Jones, D.E.: Variations in the polarity distribution of the interplanetary magnetic field. J. Geophys. Res. **71**, 2831–2839 (1966).
2.8. Bartels, J.: Some problems of terrestrial magnetism and electricity. In: Terrestrial Magnetism and Electricity. (Ed. J.A. Fleming) New York: McGraw-Hill, pp. 385–433 (1939).
2.9. Snyder, C.W., Neugebauer, M.: The relation of Mariner 2 plasma data to solar phenomena. In: The Solar Wind. (Ed. R.J. Mackin and M. Neugebauer) New York: Pergamon Press, pp. 25–34 (1966).
2.10. Colburn, D.S., Sonett, C.P.: Discontinuities in the solar wind. Space Sci. Rev. **5**, 439–506 (1966).
2.11. Siscoe, G.L., Davis, L., Jr., Coleman, P.J., Jr., Smith, E.J., Jones, D.E.: Power spectra and discontinuities of the interplanetary magnetic field: Mariner 4. J. Geophys. Res. **73**, 61–82 (1968).
2.12. Burlaga, L.F., Ness, N.F.: Tangential discontinuities in the solar wind. Solar Phys. **9**, 467–477 (1969).
2.13. Hundhausen, A.J., Bame, S.J., Ness, N.F.: Solar wind thermal anisotropies: Vela 3 and Imp 3. J. Geophys. Res. **72**, 5205–5274 (1967).
2.14. Belcher, J.W., Davis, L., Jr.: Large amplitude Alfvén waves in the interplanetary medium: II. J. Geophys. Res. **76**, 3534–3563 (1971).
2.15. Burlaga, L.F.: Microscale structures in the interplanetary medium. Solar Phys. **4**, 67–92 (1968).
2.16. — Hydromagnetic waves and discontinuities in the solar wind. To be published in Space Sci. Rev. (1971).
2.17. — Ogilvie, K.W.: Magnetic and thermal pressures in the solar wind. Solar Phys. **15**, 61–71 (1970).
2.18. Scarf, F.L.: Microscopic structure of the solar wind. Space Sci. Rev. **11**, 234–270 (1970).

Chapter III

3.1. Neugebauer, M., Snyder, C. W.: Mariner 2 observations of the solar wind, 1, Average properties. J. Geophys. Res. **71**, 4469—4484 (1966).
3.2. Hundhausen, A. J.: Direct observations of solar wind particles. Space Sci. Rev. **8**, 690—749 (1968).
3.3. Coon, J. H.: Vela satellite measurements of particles in the solar wind and distant geomagnetosphere. In: Particles Trapped in the Earth's Magnetic Field (Ed. B. M. McCormac) Dordrecht, Holland: D. Reidel, pp. 231—255 (1966).
3.4. Burlaga, L. F., Ogilvie, K. W.: Heating of the solar wind. Astrophys. J. **159**, 659—670 (1970).
3.5. Hundhausen, A. J.: Solar wind properties and the state of the magnetosphere. Ann. Geophys. **26**, 427—442 (1970).
3.6. Gosling, J. T., Bame, S. J.: Solar wind speed variations 1964—1967: An autocorrelation analysis. To be published in J. Geophys. Res. (1972).
3.7. Worthing, A. G., Geffner, J.: Treatment of Experimental Data. New York: Wiley and Sons 1948.
3.8. Bame, S. J., Asbridge, J. R., Felthauser, H. E., Gilbert, H. E., Hundhausen, A. J., Smith, D. M., Strong, I. B., Sydoriak, S. J.: A compilation of Vela 3 solar wind observations, 1965 to 1967. Los Alamos Scientific Laboratory Report LA-4536, **1** (1971).
3.9. Hundhausen, A. J., Bame, S. J., Asbridge, J. R., Sydoriak, S. J.: Solar wind proton properties: Vela 3 observations from July 1965 to June 1967. J. Geophys. Res. **75**, 4643—4657 (1970).
3.10. Montgomery, M. D., Bame, S. J., Hundhausen, A. J.: Solar wind electrons: Vela 4 measurements. J. Geophys. Res. **73**, 4999—5003 (1968).
3.11. Ness, N. F., Hundhausen, A. J., Bame, S. J.: Observations of the interplanetary medium: Vela 3 and Imp 3, 1965—1967. J. Geophys. Res. **76**, 6643—6660 (1971).
3.12. Chamberlain, J. W.: On the existence of slow solutions in coronal hydrodynamics. Astrophys. J. **141**, 320—322 (1965).
3.13. Hundhausen, A. J.: Dynamics of the outer solar atmosphere. To be published in the Proceedings of the Fourth Summer Institute for Astronomy and Astrophysics, Stony Brook, 1971.
3.14. Parker, E. N.: Dynamical properties of stellar coronas and stellar winds. II. Integration of the heat-flow equation. Astrophys. J. **139**, 93—122 (1964).
3.15. Noble, L. M., Scarf, F. L.: Conductive heating of the solar wind. I. Astrophys. J. **138**, 1169—1181 (1963).
3.16. Whang, Y. C., Chang, C. C.: An inviscid model of the solar wind. J. Geophys. Res. **70**, 4175—4180 (1965).
3.17. Durney, B.: A new type of supersonic solution for the inviscid equations of the solar wind. Astrophys. J. **166**, 669—673 (1971).
3.18. Roberts, P. H., Soward, A. M.: Stellar winds and breezes. To be published in the Proceedings of the Royal Society (1972).
3.19. Parker, E. N.: Dynamical theory of the solar wind. Space Sci. Rev. **4**, 666—708 (1965).
3.20. Sturrock, P. A., Hartle, R. E.: Two-fluid model of the solar wind. Phys. Rev. Letters **16**, 628—631 (1966).
3.21. Spitzer, L.: Physics of Fully Ionized Gases. New York: Interscience Publishers 1956.
3.22. Hartle, R. E., Sturrock, P. A.: Two-fluid model of the solar wind. Astrophys. J. **151**, 1155—1170 (1968).

3.23. Blackwell, D. E.: A study of the outer solar corona from a high altitude aircraft at the eclipse of 1954 June 30. M.N.R.A.S. **116**, 57—68 (1956).

3.24. Michard, R.: Densities electroniques dans la couronne externe du 25 fevrier 1952. Ann. Astrophys. **17**, 429—442 (1954).

3.25. Hundhausen, A. J.: Composition and dynamics of the solar wind plasma. Rev. Geophys. **8**, 729—811 (1970).

3.26. Holzer, T. E., Axford, W. I.: The theory of stellar winds and related flows. Ann. Rev. Astronomy Astrophys. **8**, 31—60 (1970).

3.27. Hartle, R. E., Barnes, A.: Nonthermal heating in the two-fluid solar wind model. J. Geophys. Res. **75**, 6915—6931 (1970).

3.28. Kuperus, M.: The heating of the solar corona. Space Sci. Rev. **9**, 713—739 (1969).

3.29. Parker, E. N.: Theoretical studies of the solar wind phenomenon. Space Sci. Rev. **9**, 325—360 (1969).

3.30. Kuperus, M.: Structure and dynamics of the solar corona. To be published in Solar-Terrestrial Physics, 1970. Dordrecht, Holland: D. Reidel 1971.

3.31. Hundhausen, A. J., Bame, S. J., Ness, N. F.: Solar wind thermal anisotropies: Vela 3 and Imp 3. J. Geophys. Res. **72**, 5265—5274 (1967).

3.32. — Nonthermal heating in the quiet solar wind. J. Geophys. Res. **74**, 5810—5813 (1969).

3.33. Barnes, A.: Collisionless heating of the solar wind plasma. II. Application of the theory of plasma heating by hydromagnetic waves. Astrophys. J. **155**, 311—321 (1969).

3.34. Osterbrock, D. E.: Heating of the solar chromosphere, plages, and corona by magnetohydrodynamic waves. Astrophys. J. **134**, 347—388 (1961).

3.35. Barnes, A., Hartle, R. E., Bredekamp, J. H.: On the energy transport in stellar winds. Astrophys. J., L53—L58 (1971).

3.36. Belcher, J. W.: Alfvénic wave pressures and the solar wind. Astrophys. J. **168**, 509—524 (1971).

3.37. Alazraki, G., Couturier, P.: Solar wind acceleration caused by the gradient of Alfvén wave pressure. Astron. Astrophys. **13**, 380—389 (1971).

3.38. Billings, D. E.: Velocity fields in a coronal region with a possible hydromagnetic interpretation. Astrophys. J. **130**, 215—220 (1959).

3.39. — Spectroscopic limitation on coronal heating mechanism. Astrophys. J. **137**, 592—600 (1963).

3.40. Weber, E. J., Davis, L., Jr.: The angular momentum of the solar wind. Astrophys. J. **148**, 217—227 (1967).

3.41. Urch, I. H.: A model of the magnetized solar wind. Solar Phys. **10**, 219—228 (1969).

3.42. Brandt, J. C., Wolff, C., Cassinelli, J. P.: Interplanetary gas. XVI. A calculation of the angular momentum of the solar wind. Astrophys. J. **156**, 1117—1124 (1969).

3.43. Whang, Y. C.: Conversion of magnetic-field energy into kinetic energy in the solar wind. Astrophys. J. **169**, 369—379 (1971).

3.44. Wolff, C. L., Brandt, J. C., Southwick, R. G.: A two-component model of the quiet solar wind with viscosity, magnetic field, and reduced heat conduction. Astrophys. J. **165**, 181—194 (1971).

3.45. Gentry, R. A., Hundhausen, A. J.: A solar wind model with magnetically inhibited heat conduction (abstract). Trans. A. G. U. **50**, 302 (1969).

3.46. Whang, Y. C., Liu, C. K., Chang, C. C.: A viscous model of the solar wind. Astrophys. J. **145**, 255—269 (1965).

3.47. Scarf, F.L., Noble, L.M.: Conductive heating of the solar wind II. The inner
       corona. Astrophys. J. **141**, 1479–1491 (1965).
3.48. Weber, E.J., Davis, L., Jr.: The effect of viscosity and anisotropy in the
       pressure on the azimuthal motion of the solar wind. J. Geophys. Res. **75**,
       2419–2428 (1970).
3.49. Scarf, F.L.: Microscopic structure of the solar wind. Space Sci. Rev. **11**,
       234–270 (1970).
3.50. Nishida, A.: Thermal state and effective collision frequency in the solar wind
       plasma. J. Geophys. Res. **74**, 5155–5157 (1969).
3.51. Cuperman, S., Harten, A.: Noncollisional coupling between the electron and
       proton components in the two-fluid model of the solar wind. Astrophys. J.
       162, 315–326 (1970).
3.52. Toichi, T.: Thermal properties of the solar wind plasma. Solar Phys. **18**,
       150–164 (1971).
3.53. Spitzer, L., Jr., Harm, R.: Transport phenomena in a completely ionized gas.
       Phys. Rev. **89**, 977–981 (1953).
3.54. Forslund, D.W.: Instabilities associated with heat conduction in the solar
       wind and their consequences. J. Geophys. Res. **75**, 17–28 (1970).
3.55. e. g., p. 42 of Present, R.D.: Kinetic Theory of Gases. New York: McGraw-
       Hill 1958.
3.56. Cuperman, S., Harten, A.: The solution of one-fluid equations with modified
       thermal conductivity for the solar wind. Cosmic Electrodynamics **1**, 205–217
       (1970).
3.57. — — Dryer, M.: Characteristics of the quiet solar wind beyond the earth's
       orbit. Preprint 1971.
3.58. — — The electron temperature in the two-component solar wind. Astro-
       phys. J. **163**, 383–391 (1971).
3.59. Newkirk, G., Jr.: Structure of the solar corona. Ann. Rev. Astronomy Astro-
       phys. **5**, 213–266 (1967).
3.60. Hewish, A.: Observations of the solar plasma using radio scattering and
       scintillation methods. In: Proceedings of the Solar Wind Conference,
       Asilomar. Pacific Grove, Calif., in press 1972.
3.61. Solar and stellar spin down: angular momentum of the solar wind, Chapter 4
       of Proceedings of the Solar Wind Conference, Asilomar. Pacific Grove,
       Calif., in press 1972.
3.62. Parker, E.N.: Dynamics of the interplanetary gas and magnetic fields. Astro-
       phys. J. **128**, 664–675 (1958).
3.63. Modisette, J.L.: Solar wind induced torque on the sun. J. Geophys. Res. **72**,
       1521–1526 (1967).
3.64. Alonso-Faus, A.: Rotation of the solar wind plasma. Planetary Space Sci.
       16, 1–6 (1968).
3.65. Schubert, G., Coleman, P.J., Jr.: The angular momentum of the solar wind.
       Astrophys. J. **153**, 943–950 (1968).
3.66. Brandt, J.C., Cassinelli, J.P.: Interplanetary gas XI. An exospheric model of
       the solar wind. Icarus **5**, 47–63 (1966).
3.67. Jockers, K.: Solar wind models based on exospheric theory. Astron. Astro-
       phys. **6**, 219–239 (1970).
3.68. Hollweg, J.V.: Collisionless solar wind. 1. Constant electron temperature.
       J. Geophys. Res. **75**, 2403–2418 (1970).
3.69. — Collisionless solar wind. 2. Variable electron temperature. J. Geophys.
       Res. **76**, 7491–7502 (1971).

3.70. Dessler, A.J.: General applicability of solar wind and solar breeze theories. Comments Astrophys. Space Sci. **1**, 31–34 (1969).
3.71. Eviatar, A., Schulz, M.: Ion-temperature anisotropies and the structure of the solar wind. Planetary Space Sci. **18**, 312–332 (1970).
3.72. Schulz, M., Eviatar, A.: Electron-temperature asymmetry and the structure of the solar wind. To be published in Cosmic Electrodynamics (1971).

Chapter IV

4.1. Neugebauer, M., Snyder, C.W.: Mariner 2 observations of the solar wind: 1. Average properties. J. Geophys. Res. **71**, 4469–4484 (1966).
4.2. Wolfe, J.H., Silva, R.W.: Explorer 14 plasma observations during the Oct. 7, 1962, geomagnetic disturbance. J. Geophys. Res. **70**, 3575–3579 (1965).
4.3. Coon, J.H.: Solar wind observations. In: Earth's Particles and Fields. (Ed. B.M. McCormac) New York, pp. 359–372 (1968).
4.4. Lazarus, A.J., Bridge, H.S., Davis, J.: Preliminary results from the Pioneer 6 MIT plasma experiment. J. Geophys, Res. **71**, 3787–3790 (1966).
4.5. Wolfe, J.H., Silva, R.W., McKibben, D.D., Mason, R.H.: The compositional, anisotropic, and nonradial flow characteristics of the solar wind. J. Geophys. Res. **71**, 3329–3335 (1966).
4.6. Hundhausen, A. J., Asbridge, J. R., Bame, S. J., Gilbert, H. E., Strong, I. B.: Vela 3 satellite observations of solar wind ions: A preliminary report. J. Geophys. Res. **72**, 87–100 (1967).
4.7. Ogilvie, K.W., Burlaga, L.F., Wilkerson, T.D.: Plasma observations on Explorer 34. J. Geophys. Res. **73**, 6809–6824 (1968).
4.8. Formisano, V., Moreno, G., Palmiotto, F.: α-particle observations in the solar wind. Solar Phys. **15**, 479–498 (1970).
4.9. Bame, S.J., Hundhausen, A.J., Asbridge, J.R., Strong, I.B.: Solar wind ion composition. Phys. Rev. Letters **20**, 393–395 (1968).
4.10. Robbins, D.E., Hundhausen, A.J., Bame, S.J.: Helium in the solar wind. J. Geophys. Res. **75**, 1178–1187 (1970).
4.11. Ogilvie, K.W., Zwally, H.J.: Hydrogen and helium velocities in the solar wind. Submitted to Solar Phys. (1971).
4.12. — Wilkerson, T.D.: Helium abundance in the solar wind. Solar Phys. **8**, 435–449 (1969).
4.13. Hirshberg, J., Bame, S.J., Robbins, D.E.: Solar flares and solar wind helium enrichments. Submitted to Solar Phys. (1971).
4.14. — Asbridge, J.R., Robbins, D.E.: Velocity and flux dependence of the solar wind helium abundance. Submitted to J. Geophys. Res. (1971).
4.15. Signer, P., Eberhardt, P., Geiss, J.: Possible determination of the solar wind composition. J. Geophys. Res. **70**, 2243–2244 (1965).
4.16. Buhler, F., Eberhardt, P., Geiss, J., Meister, J., Signer, P.: Apollo 11 solar wind composition experiment: First results. Science **166**, 1502–1503 (1969).
4.17. Geiss, J., Eberhardt, P., Buhler, F., Meister, J., Signer, P.: Apollo 11 and 12 solar wind composition experiments: Fluxes of He and Ne isotopes. J. Geophys. Res. **75**, 5972–5979 (1970).
4.18. Unsöld, A.O.J.: Stellar abundances and the origin of the elements. Science **163**, 1015–1025 (1969).
4.19. Hirayama, T.: The abundance of helium in prominences and in the chromosphere. Solar Phys. **19**, 384–400 (1971).
4.20. Sears, R.L.: Helium content and neutrino fluxes in solar models. Astrophys. J. **140**, 477–484 (1964).

4.21. Demarque, P.R., Percy, J.R.: A series of solar models. Astrophys. J. **140**, 541—543 (1964).

4.22. Weymann, R., Sears, R.L.: The depth of the convective envelope on the lower main sequence and the depletion of lithium. Astrophys. J. **142**, 174—181 (1965).

4.23. Morton, D.C.: The abundance of helium in A- and B-type stars. Astrophys. J. **151**, 285—291 (1968).

4.24. Gaustad, J.E.: The solar helium abundance. Astrophys. J. **139**, 406—408 (1964).

4.25. Biswas, S., Fichtel, C.E.: Nuclear composition and rigidity spectra of solar cosmic rays. Astrophys. J. **139**, 941—950 (1964).

4.26. Lambert, D.L.: Abundance of helium in the sun. Nature **215**, 43—44 (1967).

4.27. Durgaprasad, N., Fichtel, C.E., Guss, D.E., Reames, D.V.: Nuclear-charge spectra and energy spectra in the Sept. 2, 1966, solar-particle event. Astrophys. J. **154**, 307—315 (1968).

4.28. Iben, Icko, Jr.: The Cl^{37} solar neutrino experiment and the solar helium abundance. Ann. Phys. **54**, 164—203 (1969).

4.29. Davis, R., Jr., Harmer, S.D., Hoffman, K.C.: Search for neutrinos from the sun. Phys. Rev. Letters **20**, 1205—1209 (1968).

4.30. Parker, E.N.: Comments on coronal heating. In: The Solar Corona. (Ed. J.W. Evans) New York: Academic Press, pp. 11—19 (1963).

4.31. Yeh, Tyan: A three-fluid model of solar winds. Planetary Space Sci. **18**, 199—215 (1970).

4.32. Chandrasekhar, S.: Dynamical friction I. General considerations: the coefficient of dynamical friction. Astrophys. J. **97**, 255—273 (1943).

4.33. Geiss, J., Hirt, P., Leutwyler, H.: On acceleration and motion of ions in corona and solar wind. Solar Phys. **12**, 458—483 (1970).

4.34. Hundhausen, A.J.: Composition and dynamics of the solar wind plasma. Rev. Geophys. **8**, 729—811 (1970).

4.35. Jokipii, J.R., Effects of diffusion on the composition of the solar corona and solar wind. In: The Solar Wind. (Ed. R.J. Mackin and M. Neugebauer) New York: Pergamon Press, pp. 215—219 (1966).

4.36. DeLache, P.: Contribution a l'etude de la zone transition chromosphere et couronne. Ann. Astrophys. **30**, 827—860 (1967).

4.37. Nakada, N.P.: A study of the composition of the lower solar corona. Solar Phys. **7**, 302—320 (1969).

4.38. Seaton, M.J.: The spectrum of the solar corona. Planetary Space Sci. **12**, 55—72 (1964).

4.39. Brandt, J.C.: Chemical composition of the photosphere and corona as influenced by the solar wind. Astrophys. J. **143**, 205—266 (1966).

4.40. Holzer, T.E., Axford, W.I.: Solar wind ion composition. J. Geophys. Res. **75**, 6354—6359 (1970).

4.41. Lange, J., Scherb, F.: Ion abundance in the solar wind. J. Geophys. Res. **75**, 6350—6353 (1970).

4.42. Bame, S.J.: Spacecraft observations of the solar wind composition. In: Proceedings of the Solar Wind Conference, Asilomar. Pacific Grove, Calif., in press 1972.

4.43. — Asbridge, J.R., Hundhausen, A.J., Montgomery, M.D.: Solar wind ions: $^{56}Fe^{+8}$ to $^{56}Fe^{+12}$, $^{28}Si^{+7}$, $^{28}Si^{+8}$, $^{28}Si^{+9}$, and $^{16}O^{+6}$. J. Geophys. Res. **75**, 6360—6366 (1970).

4.44. Pottasch, S.R.: The inclusion of dielectronic recombination processes in the interpretation of the solar ultraviolet spectrum. Bull. Astron. Inst. Neth. **19**, 113–124 (1967).
4.45. Withbroe, G.L.: The chemical composition of the photosphere and the corona. In: The Menzel Symposium on Solar Physics, Atomic Spectra, and Gaseous Nebulae. (Ed. K.B. Gebbie) National Bureau of Standards Publication 353, pp. 127–148 (1971).
4.46. Billings, D.E.: A Guide to the Solar Corona. New York: Academic Press 1966.
4.47. Zirin, H.: The Solar Atmosphere. Waltham, Mass.: Blaisdell Pub. Co. 1966.
4.48. Brandt, J.C., Hodge, P.W.: Solar System Astrophysics. New York: McGraw-Hill 1964.
4.49. Delache, P.: Sur l'importance des effets de diffusion dans la couronne solaire. Compt. Rend. **261**, 643–646 (1968).
4.50. Brandt, J.C., Hunten, D.M.: On the ejection of neutral hydrogen from the sun and the terrestrial consequences. Planetary Space Sci. **14**, 95–105 (1966).
4.51. Cloutier, P.: A comment on 'The neutral hydrogen flux in the solar plasma flow', by S.-I. Akasofu. Planetary Space Sci. **14**, 809–812 (1968).
4.52. Tucker, W.H., Gould, R.J.: Radiation from a low density plasma at $10^6 - 10^8\,°K$. Astrophys. J. **144**, 244–258 (1966).
4.53. Hundhausen, A.J., Gilbert, H.E., Bame, S.J.: The state of ionization of oxygen in the solar wind. Astrophys. J. **152**, L3–L5 (1968).
4.54. — — — Ionization state of the interplanetary plasma. J. Geophys. Res. **73**, 5485–5493 (1968).
4.55. Kozlovsky, B.-Z.: The stages of ionization of oxygen and helium in the solar wind. Solar Phys. **5**, 410–416 (1968).
4.56. Tinsley, B.A.: Extraterrestrial Lyman alpha. Rev. Geophys. **9**, 89–102 (1971).
4.57. Holzer, T.E.: The interaction between interstellar helium and the solar wind. J. Geophys. Res. **76**, 6965–6970 (1971).
4.58. Banks, P.M.: Interplanetary hydrogen and helium from cosmic dust and the solar wind. J. Geophys. Res. **76**, 4341–4348 (1971).
4.59. Axford, W.I.: The interaction of the solar wind with the interstellar medium. In: Proceedings of the Solar Wind Conference, Asilomar. Pacific Grove, Calif., in press 1972.

Chapter V

5.1. Neugebauer, M., Snyder, C.W.: Mariner 2 observations of the solar wind, 1. Average properties. J. Geophys. Res. **71**, 4469–4484 (1966).
5.2. Burlaga, L.F., Ogilvie, K.W.: Heating of the solar wind. Astrophys. J. **159**, 659–670 (1970).
5.3. — — Fairfield, D.H., Montgomery, M.D., Bame, S.J.: Energy transfer at colliding streams in the solar wind. Astrophys. J. **164**, 137–149 (1971).
5.4. Siscoe, G.L., Goldstein, B., Lazarus, A.J.: An east-west asymmetry in the solar wind velocity. J. Geophys. Res. **74**, 1759–1962 (1969).
5.5. Ness, N.F., Hundhausen, A.J., Bame, S.J.: Observations of the interplanetary medium: Vela 3 and Imp 3, 1965–1967. J. Geophys. Res. **76**, 6643–6660 (1971).
5.6. Gosling, J.T., Bame, S.J.: Solar wind speed variations: An autocorrelation analysis. To be published in J. Geophys. Res. (1972).
5.7. Siscoe, G.L.: Structure and orientations of solar wind interaction fronts: Pioneer 6. To be published in J. Geophys. Res. (1972).

5.8. Gosling, J.T.: Variations in the solar wind speed along the earth's orbit. Solar Phys. **17**, 499–508 (1971).

5.9. Hundhausen, A.J., Bame, S.J., Montgomery, M.D.: Variations of solar wind plasma properties: Vela observations of a possible heliographic latitude dependence. J. Geophys. Res. **76**, 5145–5154 (1971).

5.10. Wilcox, J.M., Ness, N.F.: Quasi-stationary corotating structure in the interplanetary medium. J. Geophys. Res. **70**, 5793–5805 (1965).

5.11. — — Solar source of the interplanetary sector structure. Solar Phys. **1**, 437–445 (1967).

5.12. — The interplanetary magnetic field. Solar origin and terrestrial effects. Space Sci. Rev. **8**, 258–328 (1968).

5.13. Coleman, P.J., Jr., Davis, L., Jr., Smith, E.J., Jones, D.E.: Variations in the polarity distribution of the interplanetary magnetic field. J. Geophys. Res. **71**, 2831–2839 (1966).

5.14. Wilcox, J.M., Colburn, D.S.: Interplanetary sector structure at solar maximum. To be published in J. Geophys. Res. (1972).

5.15. Carovillano, R.L., Siscoe, G.L.: Corotating structure in the solar wind. Solar Phys. **8**, 401–414 (1969).

5.16. Parker, E.N.: Interplanetary Dynamical Processes. New York: Interscience Publishers 1963.

5.17. Sarahbi, V.: Some consequences of nonuniformity of solar wind velocity. J. Geophys. Res. **68**, 1555–1557 (1963).

5.18. Dessler, A.J.: Solar wind and interplanetary magnetic field. Rev. Geophys. **5**, 1–41 (1967).

5.19. Siscoe, G.L, Finley, L.T.: Meridional (north-south) motions of the solar wind. Solar Phys. **9**, 452–466 (1969).

5.20. — — Solar wind structure determined by corotating coronal inhomogeneities, 1. Velocity-driven perturbations. J. Geophys. Res. **75**, 1817–1825 (1970).

5.21. — — Solar wind structure determined by corotating coronal inhomogeneities, 2. Arbitrary perturbations. UCLA Dept. of Meteorology. Preprint 1971.

5.22. Goldstein, B.: Nonlinear corotating solar wind structure. Submitted to J. Geophys. Res. (1971).

5.23. Matsuda, T., Sakurai, T.: Dynamics of the azimuthally dependent solar wind. Preprint 1971.

5.24. Hundhausen, A.J.: Interplanetary shock waves and the structure of solar wind disturbances. In: Proceedings of the Solar Wind Conference, Asilomar. Pacific Grove, Calif., in press 1972.

5.25. Ogilvie, K.W.: Co-rotation shock structures. In: Proceedings of the Solar Wind Conference, Asilomar. Pacific Grove, Calif., in press 1972.

5.26. Hundhausen, A.J., Montgomery, M.D.: Heat conduction and nonsteady phenomena in the solar wind. J. Geophys. Res. **76**, 2236–2244 (1971).

5.27. Pneuman, G.W.: Interaction of the solar wind with a large scale solar magnetic field. Astrophys. J. **145**, 242–254 (1966).

5.28. — Some general properties of helmeted coronal streamers. Solar Phys. **3**, 578–579 (1968).

5.29. — Coronal streamers II: Open streamer configurations. Solar Phys. **6**, 255–275 (1969).

5.30. — Kopp, R.A.: Coronal streamers III. Energy transport in streamer and interstreamer regions. Solar Phys. **13**, 176–193 (1970).

5.31. — — Gas-magnetic field interactions in the solar corona. Solar Phys. **18**, 258—270 (1971).

5.32. Schatten, K. H., Wilcox, J. M., Ness, N. F.: A model of interplanetary and coronal magnetic fields. Solar Phys. **6**, 442—455 (1969).

5.33. Altschuler, M. D., Newkirk, G., Jr.: Magnetic fields and the structure of the solar corona I: Methods of calculating coronal fields. Solar Phys. **9**, 131—149 (1969).

5.34. Newkirk, G., Jr.: Coronal magnetic fields and the solar wind. In: Proceedings of the Solar Wind Conference, Asilomar. Pacific Grove, Calif., in press 1972.

5.35. Babcock, H. W.: The solar magnetograph. Astrophys. J. **118**, 387—396 (1953).

5.36. Schatten, K. H.: Large-scale configuration of the coronal and interplanetary magnetic field. Berkeley: Space Sciences Laboratory, University of California, Technical Report, Series 9, Issue 39 (1968).

5.37. Strong, I. B., Asbridge, J. R., Bame, S. J., Heckman, H. H., Hundhausen, A. J.: Measurements of proton temperatures in the solar wind. Phys. Rev. Letters **16**, 631—633 (1966).

5.38. Coon, J. H.: Solar wind observations. In: Earth's Particles and Fields. (Ed. B. M. McCormac) New York: Rheinhold, pp. 359—271 (1968).

5.39. Hundhausen, A. J., Bame, S. J., Asbridge, J. R., Sydoriak, S. J.: Solar wind proton properties: Vela 3 observations from July 1965 to June 1967. J. Geophys. Res. **75**, 4643—4657 (1970).

5.40. Montgomery, M. D.: The properties of solar wind electrons, May 1967—May 1968. In: Proceedings of the Solar Wind Conference, Asilomar. Pacific Grove, Calif., in press 1972.

5.41. Hartle, R. E., Barnes, A.: Nonthermal heating in the two-fluid solar wind model. J. Geophys. Res. **75**, 6915—6931 (1970).

5.42. Barnes, A., Hartle, R. E., Bredekamp, J. H.: On the energy transport in stellar winds. Astrophys. J. **166**, L53—L58 (1971).

5.43. Hundhausen, A. J.: A nonlinear, transient model of high-speed solar wind streams (abstract). EOS **52**, 915 (1971).

5.44. Parker, E. N.: Dynamical theory of the solar wind. Space Sci. Rev. **4**, 666—708 (1965).

5.45. Hartle, R. E., Sturrock, P. A.: Two-fluid model of the solar wind. Astrophys. J. **151**, 1155—1170 (1968).

5.46. Montgomery, M. D., Bame, S. J., Hundhausen, A. J.: Energy and mass content of large-scale, high-speed streams in the solar wind (abstract). EOS **52**, 915 (1971).

5.47. Bartels, J.: Some problems of terrestrial magnetism and electricity. In: Terrestrial Magnetism and Electricity. (Ed. J. A. Fleming) New York: McGraw-Hill, pp. 385—433 (1939).

5.48. Mustel', E.: Quasi-stationary emission of gases from the sun. Space Sci. Rev. **3**, 139—231 (1964).

5.49. Allen, C. W.: Relation between magnetic storms and solar activity. Monthly Notices Roy. Astron. Soc. **104**, 13—21 (1944).

5.50. Pecker, J.-C., Roberts, W. O.: Solar corpuscles responsible for geomagnetic disturbances. J. Geophys. Res. **60**, 33—44 (1955).

5.51. Billings, D. E., Roberts, W. O.: The origin of M-region geomagnetic storms. Astrophys. Norv. **9**, 147—150 (1964).

5.52. Snyder, C. W., Neugebauer, M.: The relation of Mariner-2 plasma data to solar phenomena. In: The Solar Wind. (Ed. R. W. Mackin and M. Neugebauer) New York: Pergamon Press, pp. 25—34 (1966).

5.53. Couturier, P., Leblanc, Y.: On the origin of the solar wind velocity variations. Astron. Astrophys. **7**, 254—265 (1970).
5.54. Gosling, J.T., Hansen, R.T., Bame, S.J.: Solar wind speed distributions: 1962—1970. J. Geophys. Res. **76**, 1811—1815 (1971).
5.55. Sakurai, K., Stone, R.G.: Active solar radio regions at metric frequencies and the interplanetary sector structures. Solar Phys. **19**, 247—256 (1971).
5.56. Pneuman, G.W.: private communication (1971).
5.57. Belcher, J.W., Davis, L., Jr.: Large amplitude Alfvén waves in the interplanetary medium: II. J. Geophys. Res. **76**, 3534—3563 (1971).

Chapter VI

6.1. Hundhausen, A.J.: Interplanetary shock waves and the structure of solar wind disturbances. In: Proceedings of the Solar Wind Conference, Asilomar. Pacific Grove, Calif., in press 1972.
6.2. Gold, T.: Gas Dynamics of Cosmic Clouds. (Ed. H.C. van de Hulst and J.M. Bergers) Amsterdam: North-Holland, p. 103 (1955).
6.3. Parker, E.N.: Sudden expansion of the corona following a large solar flare and the attendant magnetic field and cosmic ray effects. Astrophys. J. **133**, 1014—1033 (1961).
6.4. Colburn, D.S., Sonett, C.P.: Discontinuities in the solar wind. Space Sci. Rev. **5**, 439—506 (1966).
6.5. Petschek, H.E.: Reconnection and annihilation of magnetic fields. In: The Solar Wind. (Ed. R.J. Mackin and M. Neugebauer) New York: Pergamon Press, pp. 221—228 (1966).
6.6. The Solar Wind. (Ed. R.J. Mackin and M. Neugebauer) New York: Pergamon Press, 1966.
6.7. Sturrock, P.A., Spreiter, J.R.: Shock waves in the solar wind and geomagnetic storms. J. Geophys. Res. **70**, 5345—5351 (1965).
6.8. Hundhausen, A.J.: Shock waves in the solar wind. In: Particles and Fields in the Magnetosphere. (Ed. B.M. McCormac) Dordrecht, Holland: D. Reidel, pp. 79—81 (1970).
6.9. Ogilvie, K.W., Burlaga, L.F., Wilkerson, T.D.: Plasma observations on Explorer 34. J. Geophys. Res. **73**, 6809—6824 (1968).
6.10. Burlaga, L.F.: Hydromagnetic waves and discontinuities in the solar wind. To be published in Space Sci. Rev. (1971).
6.11. Hundhausen, A.J.: Composition and dynamics of the solar wind plasma. Rev. Geophys. **8**, 729—811 (1971).
6.12. Sonett, C.P., Colburn, D.S., Davis, L., Jr., Coleman, P.J., Jr.: Evidence for a collision-free magnetohydrodynamic shock wave in interplanetary space. Phys. Rev. Letters **13**, 153—156 (1964).
6.13. — — Briggs, B.R.: Evidence for a collision-free hydromagnetic shock wave in interplanetary space. In: The Solar Wind. (Ed. R.J. Mackin and M. Neugebauer) New York: Pergamon Press, pp. 165—175 (1966).
6.14. Gosling, J.T., Asbridge, J.R., Bame, S.J., Hundhausen, A.J., Strong, I.B.: Satellite observations of interplanetary shock waves. J. Geophys. Res. **73**, 43—50 (1968).
6.15. Taylor, H.E.: Sudden commencement associated discontinuities in the interplanetary magnetic field observed by Imp 3. Solar Phys. **6**, 320—334 (1969).
6.16. Chao, J.-K.: Interplanetary collisionless shock waves. MIT Tech. Note CSR TR-70-3. Cambridge, Mass.: Mass. Inst. of Technol. 1970.
6.17. Lazarus, A.J., Binsack, J.H.: Observations of the interplanetary plasma subsequent to the July 7, 1966, proton flare. Ann. IQSY **3**, 378—385 (1970).

6.18. Bame, S. J., Asbridge, J. R., Hundhausen, A. J., Strong, I. B.: Solar wind and magnetosheath observations during the Jan. 13–14, 1967, geomagnetic storm. J. Geophys. Res. **73**, 5761–5768 (1968).

6.19. Hirshberg, J., Alksne, A., Colburn, D. S., Bame, S. J., Hundhausen, A. J.: Observation of a solar-flare-induced interplanetary shock and helium-enriched driver gas. J. Geophys. Res. **75**, 1–16 (1970).

6.20. Hones, E. W.: Experimental observations in the magnetotail during an interplanetary disturbance. In: Intercorrelated Observations Related to Solar Events. (Ed. V. Manno and D. E. Page) Dordrecht, Holland: D. Reidel, pp. 299–308 (1970).

6.21. Ogilvie, K. W., Burlaga, L. F.: Hydromagnetic shocks in the solar wind. Solar Phys. **8**, 422–434 (1969).

6.22. Lazarus, A. J., Ogilvie, K. W., Burlaga, L. F.: Interplanetary shock observations by Mariner 5 and Explorer 34. Solar Phys. **13**, 232–239 (1970).

6.23. Bonetti, A., Moreno, G., Candidi, M., Egidi, A., Formisano, V., Pizzella, G.: Observations of solar wind discontinuities from February 24 to February 28, 1969. In: Intercorrelated Satellite Observations Related to Solar Events. (Ed. V. Manno and D. E. Page) Dordrecht, Holland: D. Reidel, pp. 436–447 (1970).

6.24. Hundhausen, A. J., Bame, S. J., Montgomery, M. D.: An observation of the February 26, 1969, Interplanetary shock wave. In: Intercorrelated Satellite Observations Related to Solar Events. (Ed. V. Manno and D. E. Page) Dordrecht, Holland: D. Reidel, pp. 567–700 (1970).

6.25. Hudson, P. D.: Discontinuities in an anisotropic plasma and their identification in the solar wind. Planetary Space Sci. **18**, 1611–1622 (1970).

6.26. Neubauer, F. M.: Jump relations for shocks in an anisotropic magnetized plasma. Z. Physik **237**, 205–223 (1970).

6.27. Hirshberg, J.: The transport of flare plasma from the sun to the earth. Planetary Space Sci. **16**, 309–319 (1968).

6.28. Lepping, R. P., Argentiero, P. D.: Single spacecraft method of estimating shock normals. J. Geophys. Res. **76**, 4349–4359 (1971).

6.29. Van Allen, J. A., Ness, N. F.: Observed particle effects of an interplanetary shock wave of July 8, 1966. J. Geophys. Res. **72**, 935–942 (1967).

6.30. Greenstadt, E. W., Green, I. B., Inouye, G. T., Sonett, C. P.: The oblique shock of the proton flare of July 7, 1965. Planetary Space Sci. **18**, 333–347 (1970).

6.31. Hirshberg, J., Bame, S. J., Hundhausen, A. J.: A solar flare disturbance in the interplanetary medium. Trans. Am. Geophys. Union **49**, 728 (1968).

6.32. Schatten, K. H., Schatten, J. E.: Magnetic field structure in solar wind disturbances. Submitted to J. Geophys. Res. (1971).

6.33. Hirshberg, J., Colburn, D. S.: Interplanetary magnetic field and geomagnetic variations—a unified view. Planetary Space Sci. **17**, 1183–1206 (1969).

6.34. Burlaga, L. F., Ogilvie, K. W.: Causes of sudden commencements and sudden impulses. J. Geophys. Res. **74**, 2815–2825 (1969).

6.35. Hundhausen, A. J., Bame, S. J., Montgomery, M. D.: Large-scale characteristics of flare-associated solar wind disturbances. J. Geophys. Res. **75**, 4631–4642 (1970).

6.36. Burlaga, L. F.: A reverse hydromagnetic shock in the solar wind. Cosmic Elect. **1**, 233–238 (1970).

6.37. — Chao, J. K.: Reverse and forward slow shocks in the solar wind. J. Geophys. Res. **76**, 7516–7521 (1971).

6.38. Chao, J.K., Formisano, V., Hedgecock, P.C.: Shock pair formation. In: Proceedings of the Solar Wind Conference, Asilomar. Pacific Grove, Calif., in press 1972.

6.39. Gosling, J.T., Asbridge, J.R., Bame, S.J., Hundhausen, A.J., Strong, I.B.: Measurements of the interplanetary solar wind during the large geomagnetic storm of April 17–18, 1965. J. Geophys. Res. **72**, 1813–1822 (1967).

6.40. Ogilvie, K.W., Wilkerson, T.D.: Helium abundance in the solar wind. Solar Phys. **8**, 435–449 (1969).

6.41. Formisano, V., Moreno, G., Palmiotto, F.: α-particle observations in the solar wind. Solar Phys. **15**, 479–498 (1970).

6.42. Hirshberg, J., Bame, S.J., Robbins, D.E.: Solar flares and solar wind helium enrichments. To be published in Solar Phys. (1972).

6.43. Robbins, D.E., Hundhausen, A.J., Bame, S.J.: Helium in the solar wind. J. Geophys. Res. **75**, 1178–1187 (1970).

6.44. Geiss, J., Hirt, D., Leutwyler, H.: On acceleration and motion of ions in corona and solar wind. Solar Phys. **12**, 458–483 (1970).

6.45. Servernyi, A.B., Shabanskii, V.P.: The generation of cosmic rays in flares. Soviet Astron. AJ (English Transl.) **4**, 583–589 (1961).

6.46. Libby, L.M.: Private communication (1971).

6.47. Bell, Barbara: Major flares and geomagnetic activity. Smithsonian Contrib. Astrophys. **5**, 69–83 (1961).

6.48. Ballif, J.R., Jones, D.E.: Flares, Forbush decreases, and geomagnetic storms. J. Geophys. Res. **74**, 3499–3511 (1969).

6.49. Solar-Geophysical Data. Washington, D.C.: U.S. Govt. Printing Office 1965 and 1967.

6.50. Kundu, M.R.: Solar Radio Astronomy. New York: Interscience Publishers 1965.

6.51. I.A.U. Quarterly Bulletin on Solar Activity. Zürich: Eidgen. Sternwarte 1967.

6.52. Akasofu, S.-I., Yoshida, S.: The structure of solar plasma flow generated by solar flares. Planetary Space Sci. **15**, 39–47 (1967).

6.53. Courant, R., Friedrichs, K.O.: Supersonic Flow and Shock Waves. New York: Interscience Publishers 1948.

6.54. Rodgers, M.H.: Analytic solutions of the blast-wave problem with an atmosphere of varying density. Astrophys. J. **125**, 478–493 (1957).

6.55. Simon, M., Axford, W.I.: Shock waves in the interplanetary medium. Planetary Space Sci. **14**, 901–908 (1966).

6.56. Lee, T.S., Balwanz, W.W.: Singular variations near the contact discontinuity in the theory of interplanetary blast waves. Solar Phys. **4**, 240–258 (1968).

6.57. — Chen, T.: Hydromagnetic interplanetary shock waves. Planetary Space Sci. **16**, 1483–1562 (1968).

6.58. — — Balwanz, W.W.: Hydromagnetic theory for disturbances following an ideal "solar thermal explosion" (abstract). EOS **51**, 414 (1970).

6.59. Korobeinikov, V.P.: On the gas flow due to solar flares. Solar Phys. **7**, 463–469 (1969).

6.60. Dryer, M.: Some effects of finite electrical conductivity on solar flare-induced interplanetary shock waves. Cosmic Elect. **1**, 348–370 (1970).

6.61. — Interplanetary double-shock ensembles with anomalous electrical conductivity. In: Proceedings of the Solar Wind Conference, Asilomar. Pacific Grove, Calif., in press 1972.

6.62. Hundhausen, A.J., Gentry, R.A.: Numerical simulation of flare-generated disturbances in the solar wind. J. Geophys. Res. **74**, 2908–2918 (1969).

6.63. — — The effects of solar flare duration on a double shock pair at 1 AU. J. Geophys. Res. **74**, 6229–6238 (1969).

6.64. DeYoung, D. S., Hundhausen, A. J.: Non spherical propagation of a flare-associated interplanetary blast wave. J. Geophys. Res. **76**, 2245–2253 (1971).

6.65. Parker, E. N.: Interplanetary Dynamical Processes. New York: Interscience Publishers 1963.

6.66. Hundhausen, A. J., Montgomery, M. D.: Heat conduction and non-steady phenomena in the solar wind. J. Geophys. Res. **76**, 2236–2244 (1971).

6.67. Dryer, M., Jones, D. L.: Energy deposition in the solar wind by flare-generated shock waves. J. Geophys. Res. **73**, 4875–4882 (1968).

6.68. Montgomery, M. D., Bame, S. J., Hundhausen, A. J.: Energy and mass content of large-scale, high-speed streams in the solar wind (abstract). EOS **52**, 915 (1971).

6.69. Bruzek, A.: Physics of solar flares, the energy and mass problem. In: Solar Physics. (Ed. J. N. Xanthakis) New York: Interscience Publishers, pp. 399–421 (1967).

6.70. Hansen, R. T., Garcia, C. J., Grognard, R. J.-M., Sheridan, K. V.: A coronal disturbance observed simultaneously with a white light coronameter and the 80 MHz Culgoora radioheliograph. Proc. A.S.A. **2**, 57–60 (1971).

6.71. Valdez, Jesusa, Altschuler, M. D.: Preliminary observations of coronal magnetic fields before and after solar proton events. Solar Phys. **15**, 446–452 (1970).

6.72. Najita, K., Orrall, F. Q.: White light events as photospheric flares. Solar Phys. **15**, 176–194 (1970).

Chapter VII

7.1. Brandt, J. C.: Introduction to the Solar Wind. San Francisco: W. H. Freeman and Company 1970.

7.2. Bernstein, W.: The solar plasma—its detection, measurement, and significance. In: Space Physics. (Ed. D. P. LeGalley and A. Rosen) New York: John Wiley and Sons, pp. 397–436 (1964).

7.3. Hundhausen, A. J.: Direct observations of solar wind particles. Space Sci. Rev. **8**, 690–749 (1968).

7.4. Vasyliunas, V. M.: Deep Plasma Measurements. In: Methods of Experimental Physics 9 (Part B). New York: Academic Press, pp. 49–89 (1971).

7.5. Ness, N. F.: Magnetometers for space research. Space Sci. Rev. **11**, 459–554 (1970).

7.6. Lust, R.: The properties of interplanetary space. In: Solar-Terrestrial Physics. (Ed. J. W. King and W. S. Newman) London: Academic Press, pp. 1–44 (1967).

7.7. Axford, W. I.: Observations of the interplanetary plasma. Space Sci. Rev. **8**, 331–365 (1968).

7.8. Vitkevich, V. V.: Scattering and scintillations of discrete radio sources as a measure of the interplanetary plasma irregularities. In: Interplanetary Medium. Part II of Solar-Terrestrial Physics/1970. (Ed. E. R. Dyer, J. G. Roederer, and A. J. Hundhausen) Dordrecht, Holland: D. Reidel, pp. 49–68 (1972).

7.9. Hewish, A.: Observations of the solar plasma using radio scattering and scintillation methods. In: Proceedings of the Solar Wind Conference, Alisomar. Pacific Grove, Calif., in press 1972.

7.10. James, J. C.: Radar studies of the sun. In: Radar Astronomy. (Ed. J V. Evans and Tor Hagfors) New York: McGraw-Hill, pp. 323–385 (1968).

7.11. Burlaga, L. F.: Hydromagnetic waves and discontinuities in the solar wind. To be published in Space Sci. Rev. (1971).

7.12. — On the nature and origin of directional discontinuities. J. Geophys. Res. **76**, 4360–4365 (1971).

7.13. Turner, J. M., Siscoe, G. L.: Orientations of "rotational" and "tangential" discontinuities in the solar wind. J. Geophys. Res. **76**, 1816–1822 (1971).

7.14. Belcher, J. W., Davis, L., Jr.: Large Amplitude Alfvén waves in the interplanetary medium: II. J. Geophys. Res. **76**, 3534–3563 (1971).

7.15. Scarf, F. L.: Microscopic structure of the solar wind. Space Sci. Rev. **11**, 234–270 (1970).

7.16. Parker, E. N.: Interplanetary Dynamical Processes. New York: Interscience Publishers 1963.

7.17. Holzer, T. E., Axford, W. I.: The theory of stellar winds and related flows. Ann. Rev. Astronomy Astrophys. **8**, 31–60 (1970).

7.18. Banks, P. M., Holzer, T. E.: High-latitude plasma transport: the polar wind. J. Geophys. Res. **74**, 6317–6332 (1969).

7.19. Axford, W. I.: The interaction of the solar wind with the interstellar medium. In: Proceedings of the Solar Wind Conference, Asilomar. Pacific Grove, Calif., in press 1972.

7.20. — Energetic solar particles in the interplanetary medium. In: The Interplanetary Medium. Part II of Solar-Terrestrial Physics/1970. (Ed. E. R. Dyer, J. G. Roederer, and A. J. Hundhausen) Dordrecht, Holland: D. Reidel, pp. 110–134 (1972).

7.21. Jokipii, J. R.: Propagation of Cosmic Rays in the Solar Wind. Rev. Geophys. **9**, 27–88 (1971).

7.22. Ness, N. F.: Interaction of the solar wind with the moon. In: The Interplanetary Medium. Part II of Solar-Terrestrial Physics/1970. (Ed. E. R. Dyer, J. G. Roederer, and A. J. Hundhausen) Dordrecht, Holland: D. Reidel, pp. 159–205 (1972).

7.23. Spreiter, J. R., Alksne, A.: Solar-wind flow past objects in the solar system. Ann. Rev. Fluid Mechanics **2**, 313–354 (1970).

7.24. Dyer, E. R., Roederer, J. G. (Ed.): The Magnetosphere. Part III of Solar-Terrestrial Physics/1970. Dordrecht, Holland: D. Reidel 1972.

7.25. Weber, E. J.: Unique solutions of solar wind models with thermal conductivity. Solar Phys. **14**, 480–488 (1970).

7.26. Dahlberg, E.: Viscous model of solar wind flow. J. Geophys. Res. **75**, 6312–6317 (1970).

7.27. Yeh, Tyan: Temperature Profile of Solar Winds. J. Geophys. Res. **76**, 7508–7515 (1971).

7.28. Durney, B. R., Roberts, P. H.: On the theory of stellar winds. To be published in Astrophys. J. (1971).

7.29. — Werner, N. E.: On the domains of existence of the three types of supersonic solutions of the inviscid solar wind equations. To be published in Astrophys. J. (1972).

7.30. Roberts, P. H.: A transformation of the stellar wind equations. Astrophys. Letters **9**, 79–80 (1971).

7.31. Durney, B. R.: Predicted solar wind properties at the earth by one-fluid models. Submitted to J. Geophys. Res. (1972).

7.32. Rosenberg, R. L., Coleman, P. J., Jr.: Heliographic latitude dependence of the dominant polarity of the interplanetary magnetic field. J. Geophys. Res. **74**, 5611–5622 (1969).

7.33. Coleman, P. J., Rosenberg, R. L.: North-south component of the interplanetary magnetic field. J. Geophys. Res. **76**, 2917–2926 (1971).
7.34. Rosenberg, R. L.: Unified theory of the interplanetary magnetic field. Solar Phys. **15**, 72–78 (1970).
7.35. Wilcox, J. M.: Statistical significance of the proposed heliographic latitude dependence of the dominant polarity of the interplanetary magnetic field. J. Geophys. Res. **75**, 2587–2590 (1970).
7.36. Suess, S. T.: On the three dimensional solar wind. To be published in J. Geophys. Res. (1972).

Subject Index